Other Books by Joan Elma Rahn

ILLUSTRATED BY GINNY LINVILLE WINTER

Grocery Store Botany
How Plants Are Pollinated
How Plants Travel
Seeing What Plants Do
More About What Plants Do
The Metric System

Alfalfa, Beans & Clover

Joan Elma Rahn

Alfalfa, Beans & Clover

ILLUSTRATED BY

Ginny Linville Winter

Atheneum · New York

1976

To Phillip Goldman

Library of Congress Cataloging in Publication Data
Rahn, Joan Elma
Alfalfa, Beans & Clover

SUMMARY: *An introduction to the characteristics, uses, and importance of about sixty members of the bean family, one of the largest and most diverse families of flowering plants. Includes Index.*

1. Legumes—Juvenile literature. [1. Legumes]
I. Winter, Ginny Linville. II. Title.
QK495.L52R3 583'.32 76–67
ISBN 0–689–30528–1

Published simultaneously in Canada by
McClelland & Stewart, Ltd.
Manufactured in the United States of America by
Halliday Lithograph Corporation,
West Hanover, Massachusetts

Designed by Mary M. Ahern

First Edition

Contents

Alfalfa, Beans & Clover

Introduction

BEANS AND PEAS. They seem like ordinary foods. Perhaps you eat them as a vegetable once or twice a week. You might think of them along with carrots and beets and broccoli as something good for you, but something you could do without—and that might be true. If you can afford to buy meat, poultry, fish, eggs, cheese, and milk—these are animal products that supply you with protein, which you require in your daily diet—then you might never need to eat beans or peas. But many poor people throughout the world cannot afford to buy expensive animal foods. Because beans and peas are richer in protein than other vegetables, many people use them as their major source of protein.

The bean family is one of the largest plant families; it has more than 14,000 species (kinds) of plants. It includes not only plants with "bean" in their names, but many additional ones, such as peas, lentils, lupines, clovers, alfalfa, vetches, licorice, and black locust trees—just to name a few. Not all are edible by human beings. Some serve as forage (food eaten by horses, cattle, and other livestock). Others may be enjoyed for their beauty. Some of the trees in the bean family yield valuable timber, and several plants provide drugs and other medicinal products. A few members of the family are poisonous to human beings or to livestock.

No matter where you live, you probably can find several plants in this family growing near you. Yesterday as I walked to the store I saw more than twenty different members of the bean family or products made from them. White clovers and black medick grow in the lawns. My neighbors have peas, green beans, and wax beans in their vegetable garden and sweet peas and lupines in the flower garden. A lovely redbud tree stands in the center of the lawn. Honey locust trees are planted along the street, and a wisteria vine covers an arbor leading to the library. At the railroad crossing I can look in

either direction and see white sweet clovers growing on both sides of the tracks. There are a few yellow sweet clovers and tick trefoils among them. Young black locust trees have sprouted along a fence near the tracks. Where a highway crosses the main street, red clover and coronilla plants bloom all summer long even though the roadside is mowed every few weeks.

In the supermarket display of dry seeds are sacks of kidney beans, pinto beans, great northern beans, and navy beans, all of which, together with green beans and wax beans, are just varieties of the same type of bean. Near them are more sacks with split peas (the ordinary green peas or a yellow variety), lima beans, soybeans, black-eyed peas, lentils, garbanzos, and peanuts. You can also buy some of these fresh, frozen, or canned. Lima beans are an important ingredient of succotash, and navy beans are the beans of Boston baked beans. Near the check-out counter are licorice-flavored cough drops and candy bars containing peanuts. Another counter has peanut butter, and not far away is the clover honey.

Two blocks down the street, in the health food store, among seeds to be used for sprouting, I found a selection of beans including mung beans. There also were carob-flavored cookies and candies and containers of carob powder.

Depending on where you live, you may find the same plants I did, or you may find different members of the bean family. It would be impossible to discuss them all here, but perhaps the descriptions in this book will help you to identify many of the members of the family that you do find and also help you to understand their importance in our lives.

Chapter 1

How Plants Are Named

BEFORE CONSIDERING the plants of the bean family, we should discuss the classification and naming of plants. Every plant family has a scientific name given to it by botanists. In the cases of most plant families, botanists are agreed on the family names, but several names have been given to the bean family. In some botany books this family is called the Leguminosae. This name is based on the Latin word *legumen,* by which the Romans referred to the edible seeds of some plants in this family. Leguminosae is the family name used in this book.

A botanist refers to a plant in this family as a *leguminous* plant, but a farmer usually calls it a *legume*. The seeds are borne in fruits called *legumes* by botanists, but most persons just call them *pods*. To avoid confusion over the two meanings of the word *legume,* we will call the plants leguminous plants and the fruits pods.

In some botany books the family Leguminosae is called Papilionaceae (from the Latin word *papilio,* which means butterfly): this is because of the resemblance of the flowers of many leguminous plants to a butterfly. In still other books the family is called Fabaceae (from the Latin word *faba,* which means bean).

The common names for this family vary, too. It has been called the bean family, the pea family, and the pulse family. (A pulse is the edible seed of a leguminous plant—not all seeds in this family are edible, however.)

The family Leguminosae is divided into three subfamilies: Papilionoideae, Caesalpinioideae, and Mimosoideae. Of these, the Papilionoideae (pea subfamily) is the largest; its members bear the typical butterflylike flowers. The Caesalpinioideae (senna subfamily) and the Mimosoideae (mimosa subfamily) are smaller groups; their flowers are not butterflylike, but their fruits are pods.

We should consider also the naming of species of plants. Each plant family (or subfamily, if the family has subfamilies), contains groups of plants called *genera* (singular, *genus*). Most genera are divided into several *species* (the singular of which is also *species*); a few genera have just one species. The species of a genus are more closely related to each other than they are to species of other genera.

The name of a species includes both the name of its genus and its own species name, called a specific epithet. For instance, the genus to which the kidney bean belongs is *Phaseolus*, and the specific epithet for the kidney bean is *vulgaris*; so the correct name for this species is *Phaseolus vulgaris*. The scarlet runner bean, which is raised as much for the beauty of its flowers as for its edible seeds, belongs to the same genus but a different species. Its name is *Phaseolus coccineus*. The common garden pea belongs not only to a different species but to a different genus as well; its name is *Pisum sativum*.

The scientific names are in Latin and so are printed in italics. If you are writing by hand or typing, you should underline these names.

The specific epithet usually describes some characteristic of the plant—*vulgaris* means common, *coccineus* means scarlet, and *sativum* means cultivated. Occasionally the discoverer of a new species wishes to honor someone by naming the species after him. In this case, the specific epithet is the latinized form of that person's name. *Phaseolus metcalfei* was named in honor of J. K. Metcalfe. The appendix gives the meanings of the specific epithets of the plants named in this book.

The generic name is always capitalized, but the specific epithet is not, although it once was customary to capitalize specific epithets derived from proper nouns, such as names of persons.

If the specific epithet is not known or is just not used, it is indicated by "sp." after the generic name. Thus, "*Phaseolus* sp." means a species of the genus

Phaseolus. "*Phaseolus* spp." means several species of the genus.

Why do we use scientific names for plants? One reason is that the name is the same the world over (except for a few cases in which the proper classification of a species is in doubt and botanists have not yet agreed on a name). Each species has its own scientific name that may not be applied to another species. Botanists in every country know what plant is meant when its scientific name is used. Common names for a given plant, on the other hand, may be different in different places. Some plants have three, four, five, or even more common names. Furthermore, the same common name may be applied to different plants in different places. *Trifolium pratense* is a very common forage plant that frequently grows wild throughout North America as well as in Europe and Asia. Its scientific name is *Trifolium pratense* the world over, but in the United States it has the common names of red clover, purple clover, bee-bread, cock's head, honeysuckle, sugarplums, and trefoils—not to mention its names in countries where languages other than English are spoken. Of the common English names, red clover is probably the one by which this plant is best known. You may recognize honeysuckle as the name of another plant, one not even a member of the bean family. In fact, there are several plants with the common name of honeysuckle. Trefoil is part of the names of several plants, some of them members of the bean family and some of them not.

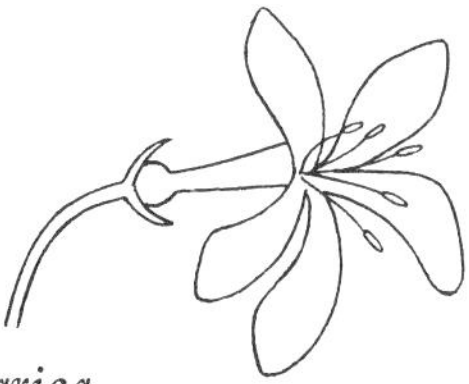

Lonicera tatarica
(HONEYSUCKLE FAMILY)
honeysuckle,
tartarian honeysuckle

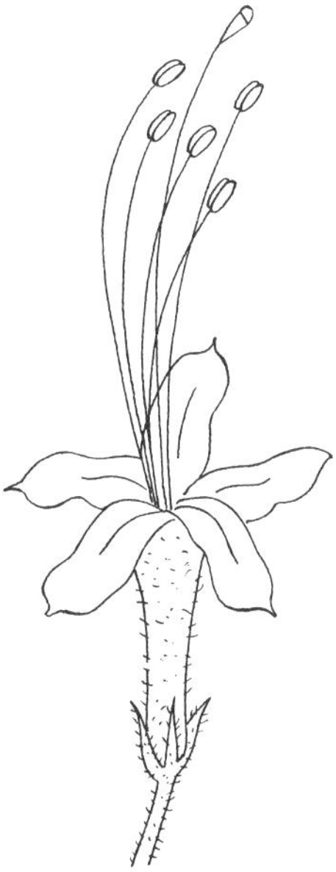

Rhododendron nudiflorum
(HEATH FAMILY) purple
azalea, pink azalea,
pinkster flower,
wild honeysuckle

Trifolium pratense
(BEAN FAMILY) red
clover, purple clover,
bee-bread, cock's head,
honeysuckle, sugarplums,
trefoils

THREE DIFFERENT PLANTS
WITH THE COMMON NAME OF HONEYSUCKLE

The word *bean* in the common name of a plant does not necessarily mean that the plant is a member of the bean family. The following are some examples: buckbean or bogbean, Indian bean, Mexican jumping bean or bean tree, castor bean, and hog's-bean. Coffee beans and cocoa beans are not true beans, either.

In this book, we will give the scientific name of each plant mentioned along with its common name (or with two or three of its best-known common names, if it has several). Some plants have no common name, but every plant thus far discovered and studied by botanists has a scientific name.

Chapter 2

The Bean Family

(*Family Leguminosae*)

A FAMILY as large as the bean family, with its 14,000 or more species, usually has a great deal of variation within it, and this is certainly true of the Leguminosae. A person seeing a small clover plant and a large royal poinciana tree for the first time probably would not think of them as being closely related. A shrubby mesquite tree growing in the dry desert lands of Arizona would at first seem to have little in common with lianas (tropical vines) climbing more than 200 feet to the treetops of the humid rain forests of southeastern Asia. Yet these plants and other members of the bean family do have certain resemblances; that is why most botanists classify them in the same family.

The family name, Leguminosae, suggests the feature that all members of the family have in common

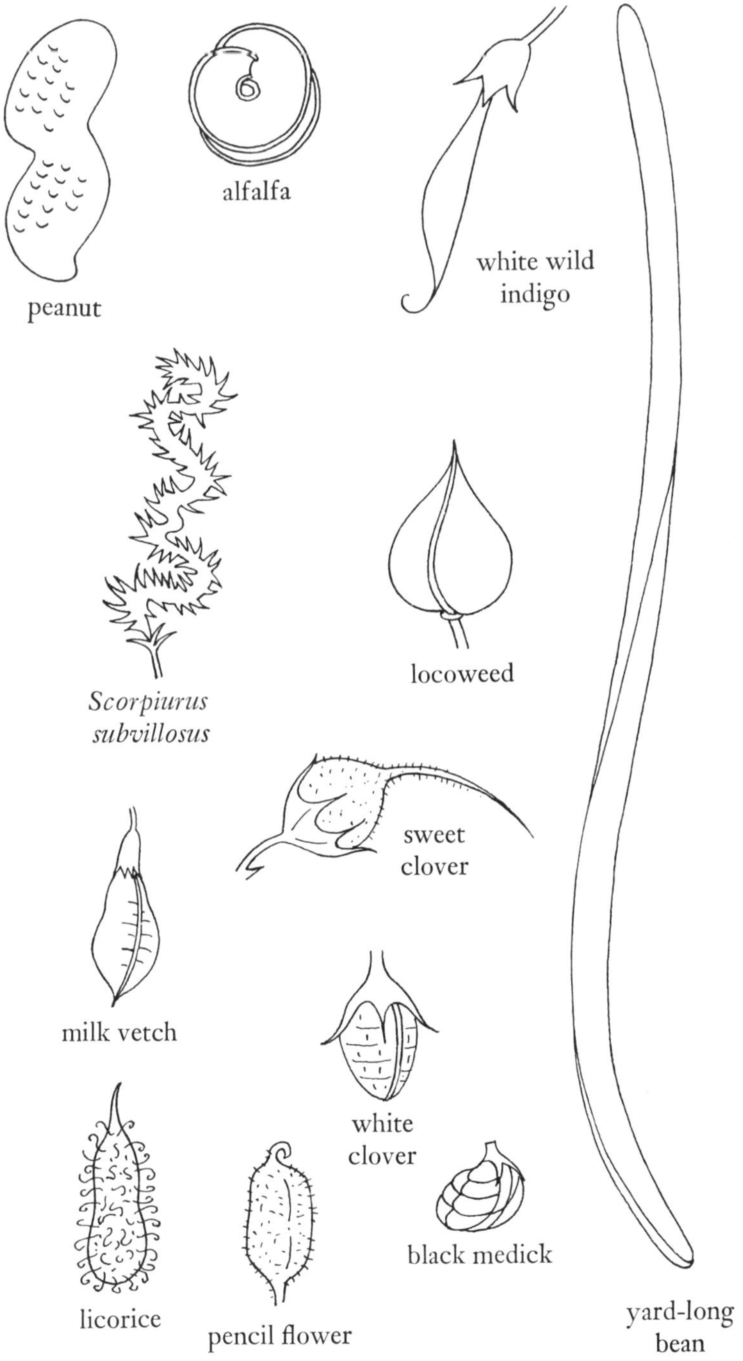
peanut
alfalfa
white wild
indigo
yard-long
bean
Scorpiurus
subvillosus
locoweed
sweet
clover
milk vetch
white
clover
licorice
pencil flower
black medick

SEVERAL PODS

—a fruit that is a pod or legume. The pods of most leguminous plants resemble those of beans and peas in shape; they are long and slender and contain a row of several seeds. There are some exceptions to this, however. Peanuts (*Arachis hypogaea*) have dry, fibrous pods, and the pods of alfalfa (*Medicago sativa*) are small and spirally twisted. The clovers (*Trifolium* spp.) have tiny, spherical pods with only one or a few seeds, but the woody pods of the gogo-vine (*Entada phaseoloides*), a huge tropical liana, may be as much as five feet long and contain numerous seeds.

GOGO-VINE POD

BIRD'S-FOOT TREFOIL

In many members of the bean family the pods dry as they ripen and then suddenly split open along opposite sides. In several of these plants—bird's-foot trefoil (*Lotus corniculatus*), for instance—the pods twist as they open thus scattering their seeds. In tropical Africa there is a tree, *Berlinia grandiflora,* which has pods about the shape and texture of the leather sole of a shoe. They are usually bigger than the sole of a shoe—as much as a foot long—and the seeds scatter as the pods open explosively with a loud noise.

In other leguminous plants—mesquite (*Prosopis* spp.), for instance—the pods do not split open, and the seeds may be carried away by animals that eat the pods. Wind and water sometimes distribute pods and seeds.

The pods of tick trefoil or beggar's-lice (*Desmodium* spp.) are divided into segments, each containing one seed. Such a pod, called a *loment,* breaks

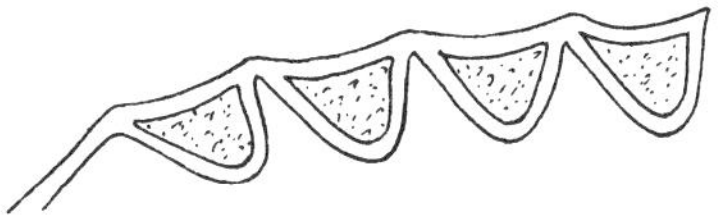

LOMENT OF TICK TREFOIL

HAIR

easily into its individual segments at the slightest touch when ripe. Fine, hooked hairs cover the segments. A passing animal brushing a plant can hardly help but break off the segments of several loments, which then become firmly attached to its fur. Only later—perhaps days or weeks later—do the segments fall to the ground as the animal sheds its fur or the fur becomes caught on shrubs. Perhaps you yourself have been annoyed to find these loments attached to your clothing after an autumn walk in the woods. So tenaciously do they stick, that, if you fail to remove them, they often remain in place while the clothing goes through the laundry.

The leaves of plants belonging to the family leguminosae may differ even more than the pods. A leaf has two parts: a flat, green portion called a *blade,* and a stalk, called the *petiole.* One end of the petiole is attached to the blade, the other to the stem from which the leaf grows. In a *simple leaf* the

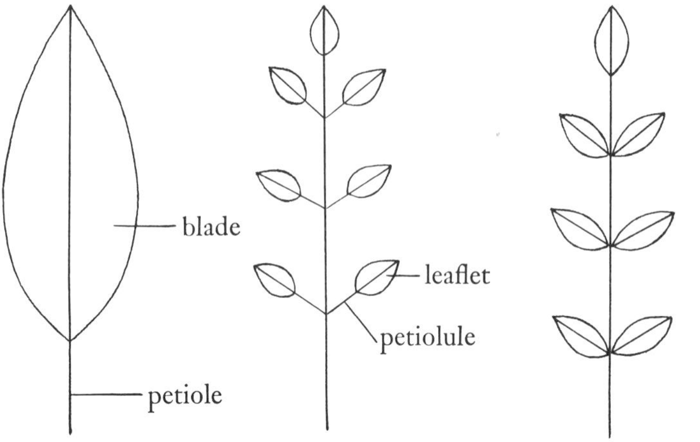

SIMPLE LEAF

COMPOUND LEAF WITH PETIOLULES

COMPOUND LEAF WITHOUT PETIOLULES

SIMPLE AND COMPOUND LEAVES

blade is all in one piece; in a *compound leaf* the blade is divided into several smaller pieces called *leaflets.* The leaves of most leguminous plants are compound, though a few species have simple leaves. Some leaflets have their own stalks, called *petiolules,* but others do not.

Sometimes you may wonder if you are looking at a compound leaf with several leaflets or a stem with several simple leaves. In that case, examine the specimen for buds. There is nearly always a bud just above the place on the stem where a leaf grows from it—that is, at the base of the petiole. Sometimes these buds are very small, and you have to look very closely to find them. A thick petiole can hide a small bud, and you might even have to break the petiole from the stem in order to see the bud. There never are buds at the bases of leaflets or their petiolules. This should be enough for you to distinguish a stem with simple leaves from a compound leaf, but you can double-check by examining the tip of the specimen. There will be a bud or cluster of young leaves at the tip of a stem, but there will never be a bud at the tip of a compound leaf (though there may be a leaflet).

There are several types of compound leaf. In a *palmately compound* leaf the leaflets all are at-

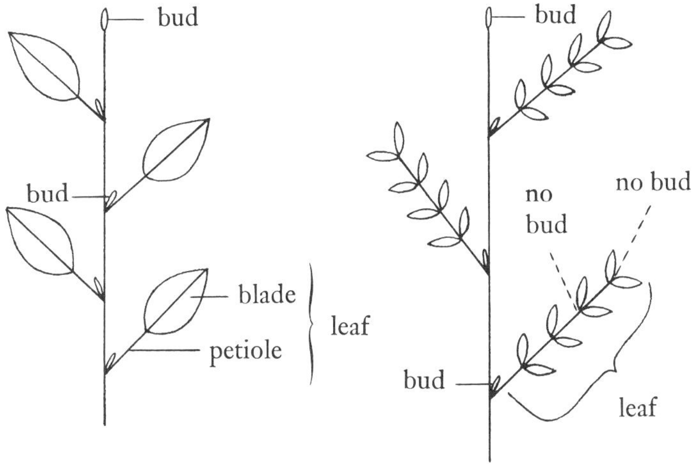

STEM WITH SIMPLE LEAVES

STEM WITH COMPOUND LEAVES

tached to the tip of the petiole; they fan out from the petiole somewhat as your fingers fan out from the palm of your hand when you spread them out as much as you can. Lupines (*Lupinus* spp.) have palmately compound leaves (page 80).

A *pinnately compound* leaf has a *rachis*, which is a continuation of the petiole; the leaflets are arranged along either side of the rachis. Sometimes there is a leaflet at the tip of the rachis and sometimes not. A leaflet at the tip is called a *terminal leaflet*; the others are *lateral leaflets*. The wisterias (*Wisteria* spp.) have pinnately compound leaves (page 83).

A *trifoliate* leaf has three leaflets at the tip of the petiole. The clovers (*Trifolium* spp.) and the sweet clovers (*Melilotus* spp.) have trifoliate leaves (pages 69 and 70).

Doubly compound leaves are compound leaves in which the leaflets are divided into even smaller leaflets. The royal poinciana (*Delonix regia*) and the acacias (*Acacia* spp.) have doubly compound leaves that give these trees a ferny appearance when they are in leaf (page 105).

Some petioles have appendages called *stipules*. The stipules are paired, one on either side of the base of the petiole. The stipules often are leafy and look like small blades, as they do in leaves of garden pea (*Pisum sativum*). Kidney bean (*Phaseolus vulgaris*) leaves have tiny, green, pointed stipules, and in the acacias and black locust (*Robinia pseudo-*

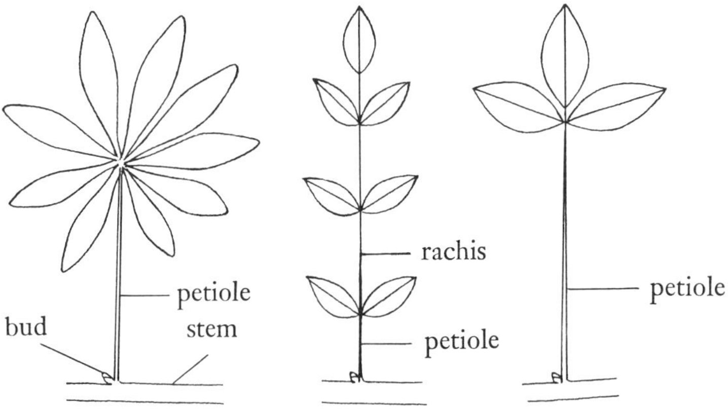

PALMATELY COMPOUND

PINNATELY COMPOUND

TRIFOLIATE

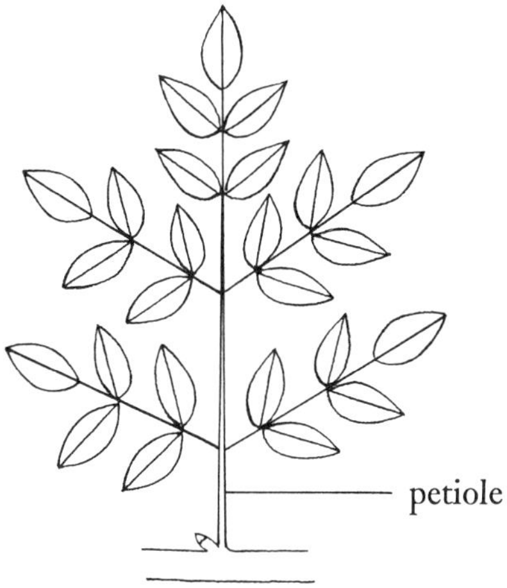

DOUBLY COMPOUND

SEVERAL TYPES OF COMPOUND LEAF

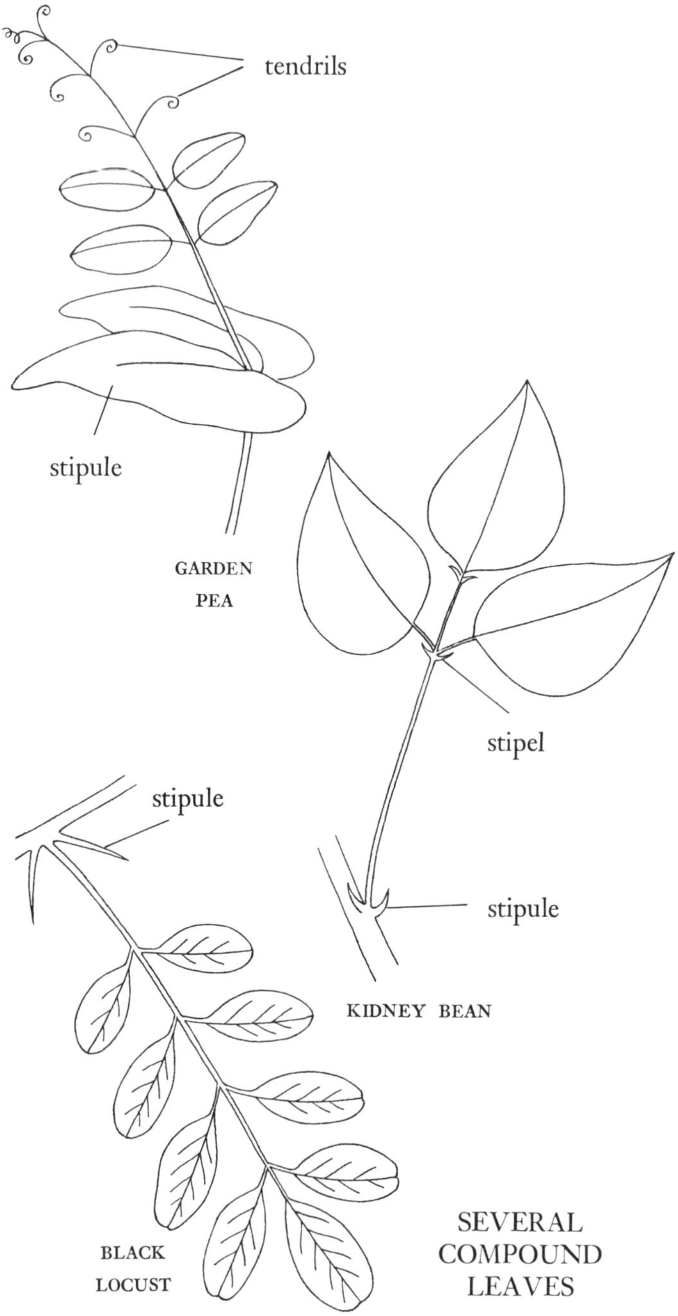

SEVERAL
COMPOUND
LEAVES

acacia) the stipules are hard, sharp, woody spines. The leaflets of some compound leaves have their own small stipules called *stipels*.

Being slender plants, vines often are not strong enough to support their own weight. They can grow tall only if there is a nearby tree, fence, or other sturdy object on which they can climb. Some part of the vine usually twines around the support. In the pole varieties of kidney beans, the stem twines, but the terminal leaflets and adjacent lateral leaflets of the garden pea are modified into twining tendrils.

The leaves of some leguminous plants (and some plants in other families, too) change their position by day and by night in what are called *sleep movements*, even though the plants, of course, do not sleep. You can observe sleep movements in the leaves of kidney bean or clover plants. The lower leaves of kidney bean plants are simple leaves. During the day, when the sun shines, the blades of these leaves are in a horizontal position (or, if the sun is low in the sky, they may be tilted in such a position that the upper surfaces of the blades face the sun). About sunset the blades bend downward, and at night they are in a vertical position. The next morning they return to the horizontal position. Other leaves of kidney beans are trifoliate; their leaflets rise and fall the same way.

Clovers have trifoliate leaves; during the day, the leaflets are in a horizontal or nearly horizontal position. About sunset the two lateral leaflets bend

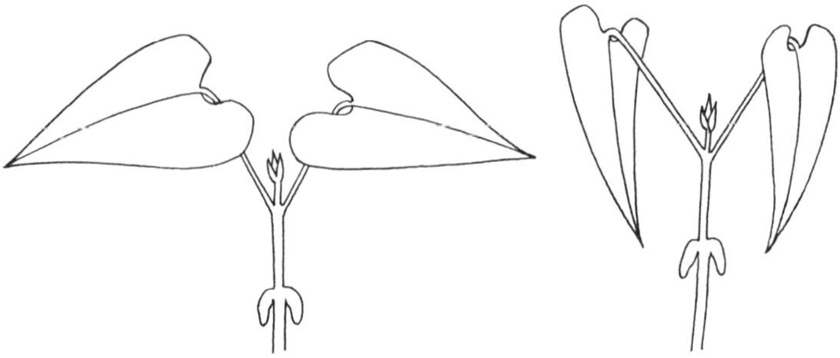

day position night position

SLEEP MOVEMENT OF BEAN LEAVES

day position night position

SLEEP MOVEMENT OF CLOVER LEAF

towards each other until their upper surfaces touch. The terminal leaflet folds down over the lateral leaflets. The next morning the leaflets return to their horizontal position.

The flowers of each subfamily are very different, and so they are not described here but in later chapters. The flowers, however, are nearly always in *in-*

florescences. An inflorescence is a cluster of flowers that grow close together on a plant. There are about a dozen different kinds of inflorescence, but only three are common in the Leguminosae: spikes, racemes, and heads. A *spike* is a long stem with many flowers growing directly from it. A *raceme* is similar to a spike, but each flower has its own short stem, which grows from the main stem. A *head* is a very short spike; in leguminous plants the heads usually are spherical or nearly so. A few leguminous plants have a fourth type of inflorescence, the *panicle,* which is a branched raceme.

Perhaps one of the most interesting and important features of leguminous plants is their protein content, which usually is much higher than that of

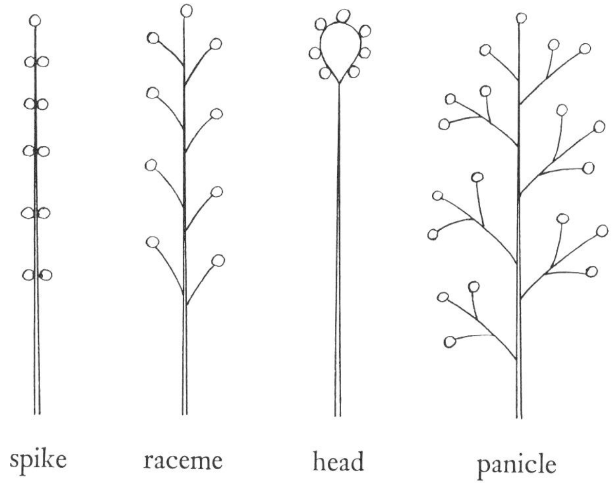

SEVERAL TYPES OF INFLORESCENCE

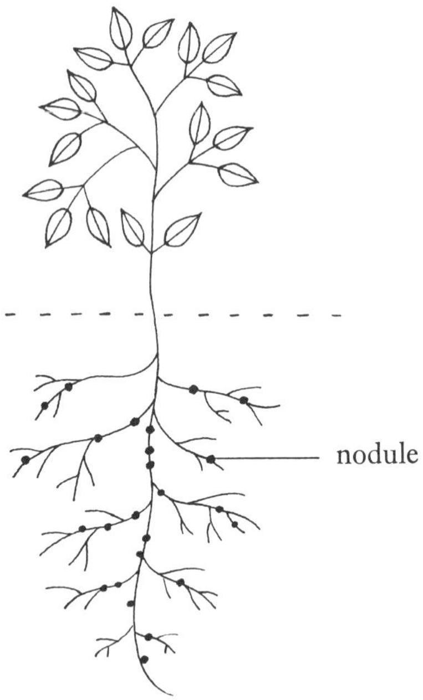

ROOT WITH ROOT NODULES

plants of other families. This is due to the ability of leguminous plants to harbor in their roots several species of bacteria called rhizobia (*Rhizobium* spp.). Rhizobia can grow and reproduce in the soil, but if leguminous plants are growing in that soil, the rhizobia may enter the roots. If they do, the roots form swellings or tumors, called *nodules.* In the root, the rhizobia live only in the nodules, but one plant may have dozens or even hundreds of nodules on its roots.

The nodules absorb nitrogen gas from the air in

the spaces among the soil particles and use it in the manufacture of several nitrogen-containing substances including amino acids, from which the plant then makes proteins. This process of utilizing nitrogen gas is called *nitrogen fixation.*

Most nodules are red inside because they contain the pigment *hemoglobin,* which is almost identical to the hemoglobin that makes our blood red. If you dig up roots with nodules large enough to cut open conveniently, you should be able to see the red color. Only the portion of the nodule that contains hemoglobin fixes nitrogen. Some nodules lack hemoglobin and do not fix nitrogen.

About 78 percent of the air consists of nitrogen gas, but most living things cannot use nitrogen in this form. They must get their nitrogen from nitrogen-containing compounds, such as amino acids, proteins, nitrates, and ammonium compounds. Animals require proteins and amino acids, which they obtain by eating other animals or plants. Most plants manufacture amino acids and proteins from the nitrates and ammonium compounds that their roots absorb from the soil; these plants have a low protein content. Only those plants that fix nitrogen have a high protein content. A few species of algae and bacteria and a few nonleguminous flowering plants that form root nodules with bacteria in them fix nitrogen, but by far the most important nitrogen fixers among crop and forage plants are leguminous plants. About 94 percent of the Papilionoideae, 34 percent of the

Caesalpinioideae, and 91 percent of the Mimosoideae fix nitrogen. It is for this reason that beans and peas are important foods for us and that alfalfa and clovers are important forage crops for farm animals.

Nitrogen-fixing leguminous plants also excrete nitrogen-containing compounds into the soil and so enrich it. In ancient times and up until the Middle Ages, farmers would let their fields lie fallow every so often, usually once in every seven years. During the fallow year they raised no crops in those fields. This was done to conserve water in the ground and to allow the stubble of the crop plants to decay thoroughly thus restoring the fertility that had gradually been lost when the land was farmed every year. Then farmers began to plant clover as a forage plant during the "fallow" year, and this was the beginning of crop rotation, which replaced fallowing.

Today a farmer who rotates crops might plant any leguminous nitrogen-fixing plant appropriate for his climate every few years in order to replace the nitrogen lost from the soil during the other years when he plants nonleguminous crops. He chooses a leguminous plant from which he may profit—either by selling a food crop such as beans or peas or by allowing livestock to graze on clover or alfalfa. Whatever use he makes of the crop, he plows under the remains of the plants at the end of the season, thus enriching the soil even more. This is called *green manuring*. Farmers who do not rotate leguminous crops with their nonleguminous crops must buy more nitrogen-rich fertilizer for their fields than farmers who do.

Leguminous plants cannot fix nitrogen without rhizobia in their roots; neither can the rhizobia do it by themselves. Only the root nodules together with rhizobia in them can fix nitrogen. To ensure that the bacteria are present in the soil, a farmer may purchase rhizobia to mix with seeds before he plants them. This way he can be sure that the bacteria will be there to enter the roots as soon as they form on the young seedlings. This is especially important if no leguminous crops have been growing in the field for many years, for then the numbers of rhizobia in the soil decline rapidly.

There are several species of *Rhizobium,* and each one grows in the roots of only one group of closely related species of leguminous plants. There are seven main groups of crop plants, each of which can be infected by a different species of *Rhizobium*: soybean group, lupine group, alfalfa group, clover group, bean group, pea and vetch group, and cowpea group.

Some of the American Indians had another way of accomplishing what European farmers did by crop rotation. The Indians grew several different kinds of vegetables, including beans, together in mixed plots. Often they planted beans next to corn, and the tall corn plants served as poles on which the bean plants could climb. While all the crops were growing, the beans helped to enrich the soil.

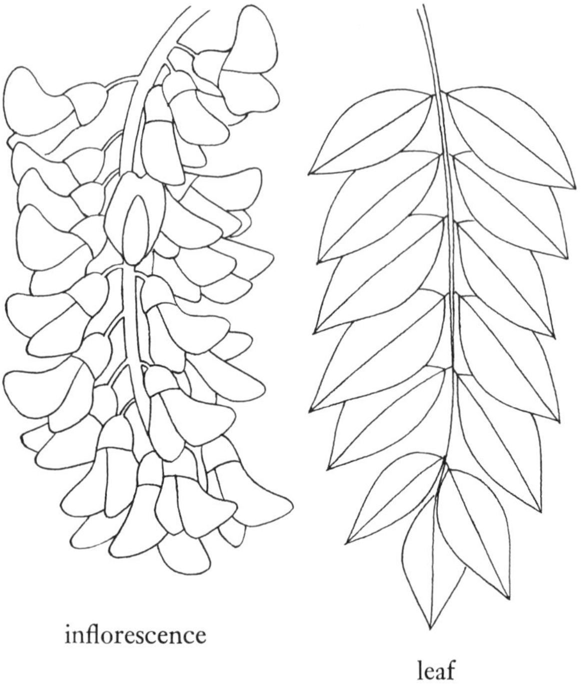

Madre de Cacao

Cacao trees, from which we obtain cocoa, are non-leguminous tropical plants that require shade. On plantations, rows of taller shade trees are planted between the rows of cacao trees. Often, leguminous trees are chosen for the shade trees, and they perform an extra service by enriching the soil with nitrogen compounds. These trees, regardless of the species, are called cacao mamma by the local people. One of them, *Gliricidia sepium,* was named *madre de cacao* (which means mother of cacao) by the

Spaniards, and it is still called by that name in English-speaking countries.

Leguminous trees often are used the same way in coffee plantations, where they are called coffee mamma.

Nitrogen fixation is also important in restoring fertility to land from which all the vegetation has been removed, as by some kinds of lumbering or strip mining. When such areas are to be reclaimed, a mixture of plants, including leguminous species are planted.

Chapter 3

The Pea Subfamily

(*Subfamily Papilionoideae*)

THE SUBFAMILY Papilionoideae is the largest of the three subfamilies in the Leguminosae, and it is the one you are thinking of when you think of beans and peas. There are at least 10,000 species in this subfamily. Many are herbs (small, nonwoody plants), but some are shrubs, trees, or vines. Except for the extremely cold polar regions covered by ice and snow the year around, nearly every land area of the world has at least a few members of this subfamily. Most of the species are tropical, and it is in the tropics that most of the trees and lianas of the pea subfamily are found. In temperate regions, most of the members are herbs and shrubs, and the vines are smaller and herbaceous (like sweet peas and the pole varieties of kidney beans); here only a few members of the subfamily are trees. Arctic regions have still fewer members, and these are mostly small plants. You can find some members of this subfamily in woods and prairies, in hot and cold climates, in tropical rain forests where the air is humid all the time, and in deserts where the air is dry and it rarely rains.

If we were to make lists of plants classified according to whether they produce food suitable for human beings, forage for animals, timber, or beautiful flowers used for decorative purposes, each list would contain several members of this subfamily. If we were to prepare a list of weeds, we would find very few members on it—perhaps the notable exceptions would be the locoweeds (*Astragalus* spp. and *Oxytropis* spp.), which are poisonous to horses, cattle, and other grazing animals, and the kudzu vine (*Pueraria lobata*), which has destroyed many trees by growing over them and shading them from the sun. The majority of leguminous plants that grow wild along roadsides and in vacant lots are species like clover or alfalfa that have escaped from cultivation; that is, they sprout from seeds accidentally carried from a farm or garden. Many of them are pretty enough to be included in wild flower books.

As diverse as plants of the pea subfamily are, there is one feature that they have in common by which you can recognize them: their flowers. Like other flowers, they have four different kinds of parts—sepals, petals, stamens, and a pistil—but the petals give the flowers their butterflylike appearance.

Each flower has five petals. One of them, ordinarily the uppermost, is the largest and also the most showy. It generally stands upright, or nearly so, and is called the *standard.* The two lowermost petals are fused together along their lower edges and form a structure called a *keel* because its shape usually resembles the keel of a boat. However, not all keels look like this. In the jade vine (*Strongylodon macrobotrys*) the keel curves upward and resembles a Turkish slipper, and in kidney beans and green beans it is twisted into a spiral (page 63). The remaining two petals—one on either side of the flower—are the *wings.* In some species the wings stand out; in other species they hug the keel, and you do not notice them until you examine the flower closely. In a few species the wings may be larger than the standard; in others, they are extremely small.

This arrangement of standard, wings, and keel is so typical of the members of the pea subfamily that you should have little difficulty recognizing them in bloom. There are only a few exceptions to this general arrangement. Leadplant and the false indigos (*Amorpha* spp.), for instance, have no wings or keel; there is only the standard, which curves around the stamens and pistil.

Five small green sepals surround the bases of the petals. The sepals are fused to each other, and together they may look like a little bell-shaped structure or a long tube. Whatever its shape, you usually

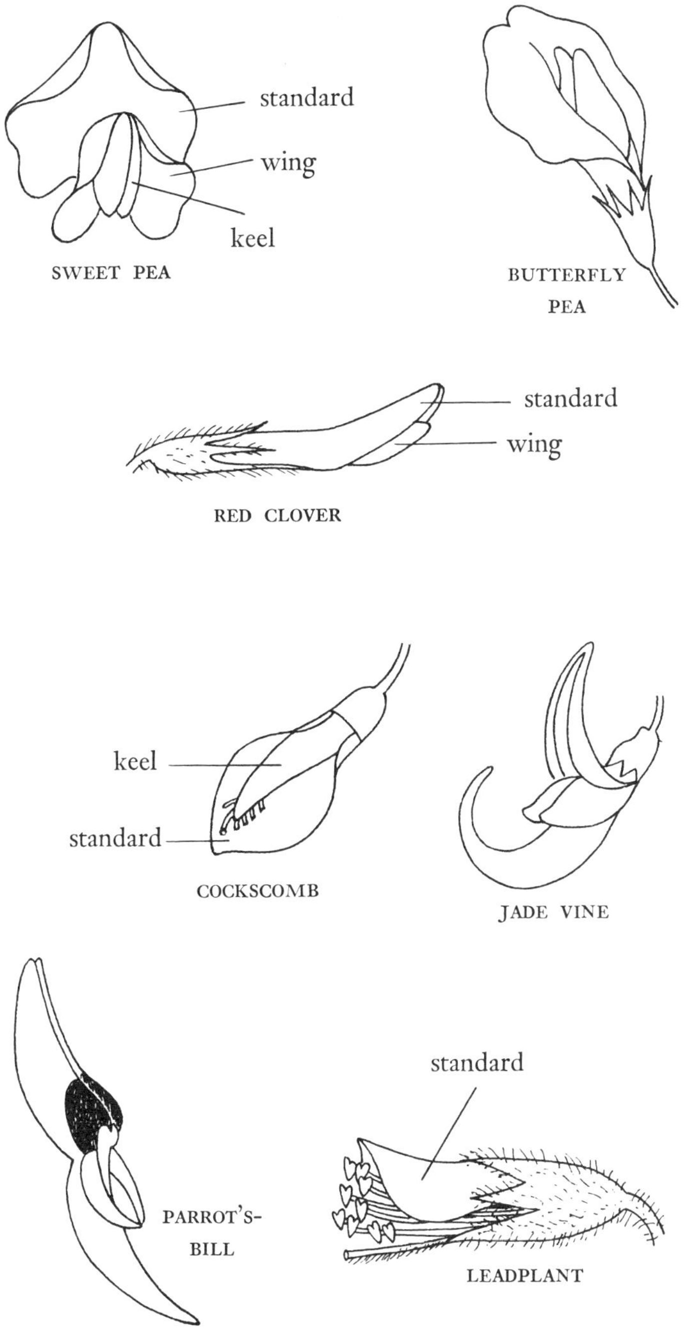

SOME VARIATIONS IN FLOWERS

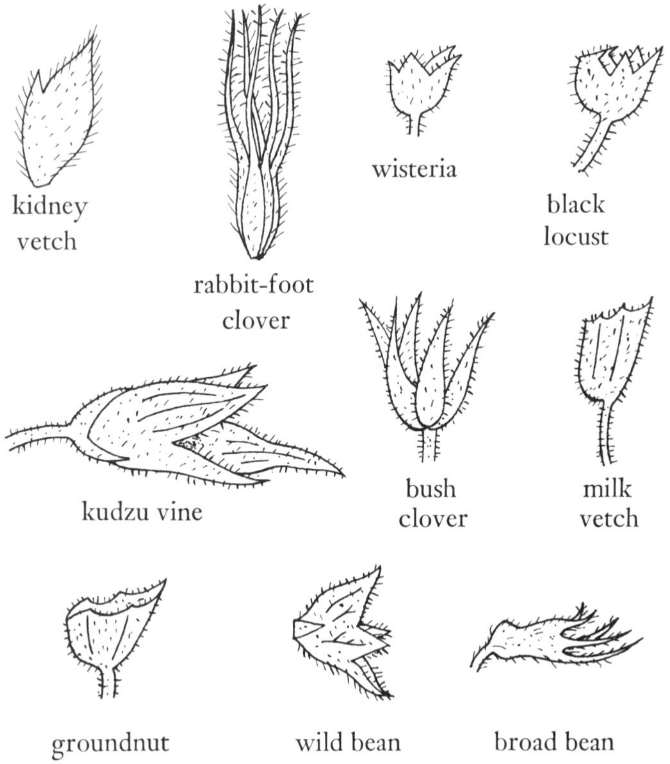

SOME VARIATIONS IN SEPALS

can tell that there are five sepals by the five projections at the top of the cup or tube. Actually the bases of all the petals are fused together, too, but that may be difficult to see without taking the flower apart.

Between the two petals of the keel are ten stamens, the male parts of the flower, and one pistil, the female part. The pistil is in the center and the ten stamens surround it.

Each stamen consists of a long stalk, called a *filament*, and an *anther* at the top of the filament. The anther produces pollen grains. When the pollen grains are ripe, the anthers open and release the pollen, which then may be transferred to the pistil of a flower. This process is called *pollination.*

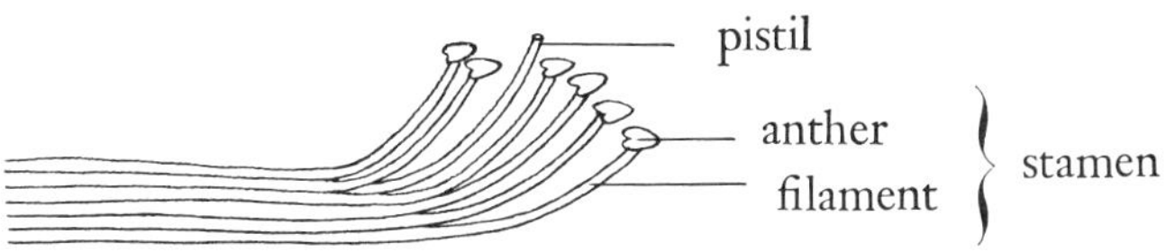

WILD INDIGO—
All Stamens Separate

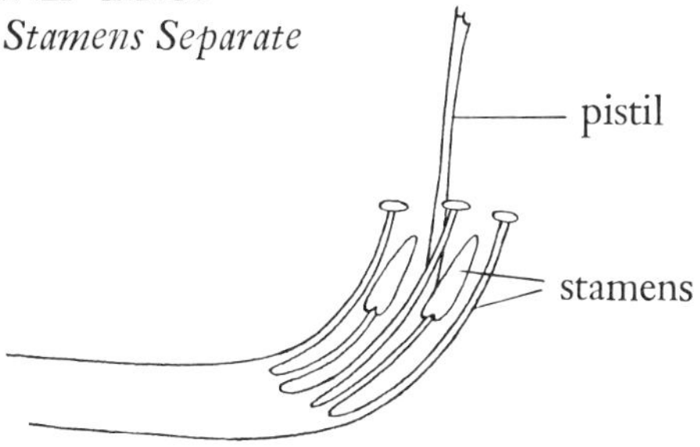

LUPINE—
All Stamens Fused

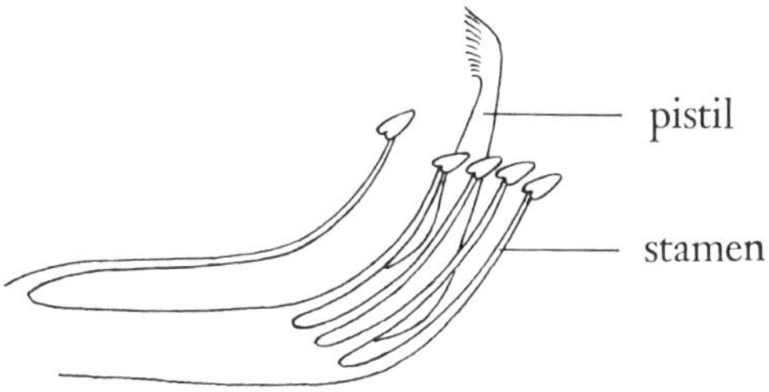

GARDEN PEA—
One Stamen Free, Others Fused

SOME ARRANGEMENTS OF STAMENS

In only a few species of the pea subfamily are all the stamens separate from each other; the wild indigos (*Baptisia* spp.) are examples. In most species, the filaments of the stamens are fused in one of two arrangements. In one arrangement the filaments of all ten stamens are fused into a tube that completely surrounds the lower portion of the pistil; lupines (*Lupinus* spp.), which are common wild flowers and garden flowers, have this arrangement. The other arrangement of stamens—by far the most common in this subfamily but rare in other plants—is one in which the upper stamen (the one nearest the standard) is free from the tube; this means that the tube is slit open on its upper side. Some examples are the stamens of kidney and lima beans (*Phaseolus* spp.), garden peas (*Pisum sativum*), peanut (*Arachis hypogaea*), alfalfa (*Medicago sativa*), clovers (*Trifolium* spp.), and wisterias (*Wisteria* spp.).

The pistil consists of three parts: ovary, style, and

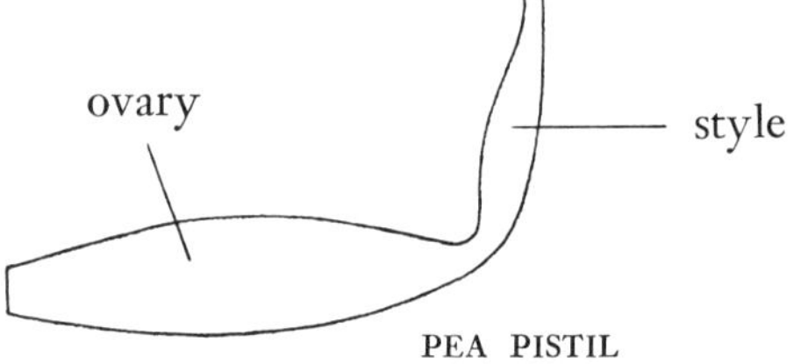

PEA PISTIL

stigma. The *ovary* is cylindrical in the pistils of most members of the bean family (but not in all flowering plants). At one end of the ovary is a slender stalk, the *style*; and the *stigma* is at the tip of the style. The ovary contains several small structures called *ovules*. If the stigma is pollinated, each of the ovules may grow into a seed; at the same time the ovule ripens into a pod.

Most flowers in the pea subfamily are pollinated by bees, who visit the flowers either to gather pollen for their young or to obtain nectar from which they make honey. Nectar is a sweet mixture of sugar and water secreted by many bee-pollinated flowers. As the bees work in the flowers, some of the pollen becomes caught among the hairs on their bodies, and when the bees visit another flower some of this pollen brushes off onto a stigma. The transfer of pollen from the stamens of one flower to the pistil of a flower on another plant is called *cross-pollination*.

Among bee-pollinated flowers of the pea subfamily there are four main types of pollination story, and we will examine one example of each: clovers, coronilla, vetches, and alfalfa.

The wings and keel of a clover flower have notches that interlock, and the wings have claws that hook over the edges of the keel. When a bee visits a clover flower, it rests its forelegs on the wings, and this presses down both the wings and the keel. Because the stamens and pistil are rather stiff and do not bend, the pressing down of the keel exposes both the anthers and the stigma. As the bee collects nectar, some pollen from the anther brushes off on the underside of the bee's head. If the bee has visited another clover flower previously, it may be carrying pollen from that flower in the same position on its head, and some of this may brush off onto the stigma. When the bee leaves, the wings and keel spring back to their original position and once more enclose the stamens and pistil. Bees making later visits repeat the same process. The sweet clovers and locoweeds have similar pollination stories.

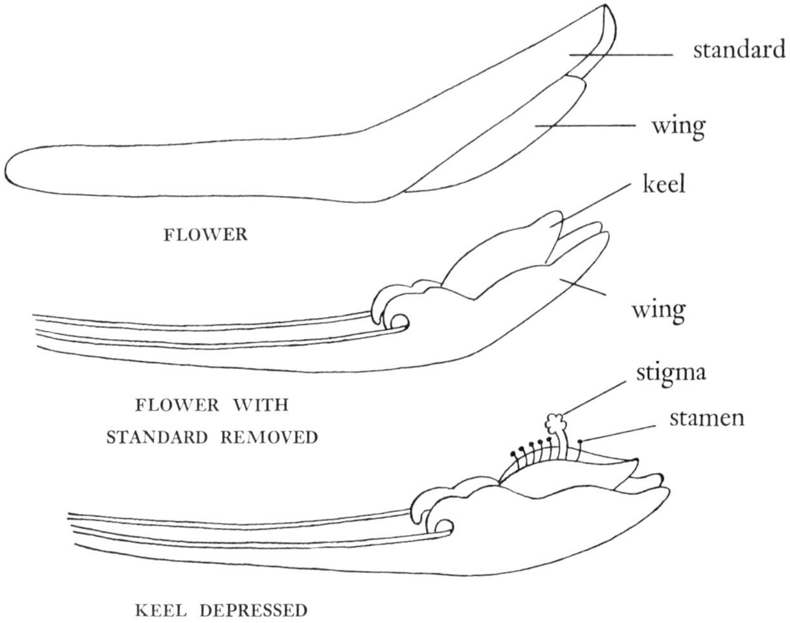

POLLINATION MECHANISM OF
RED CLOVER

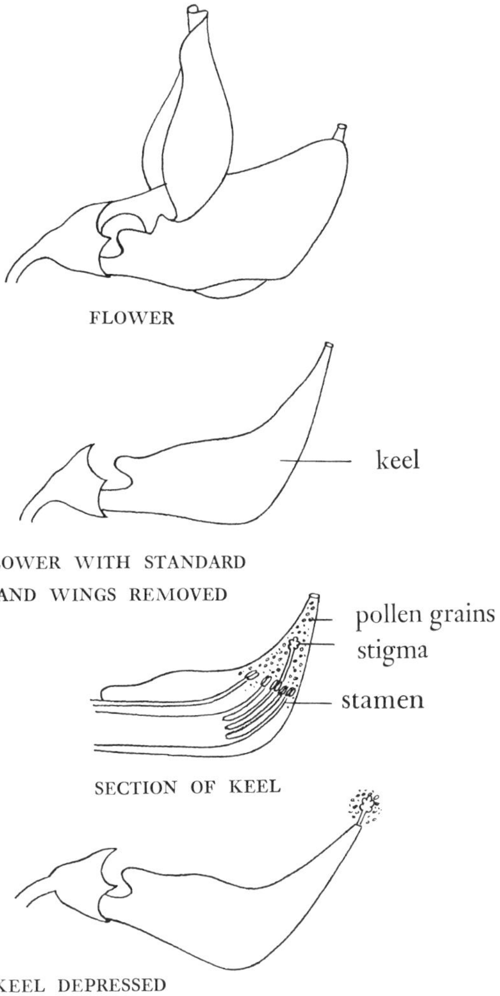

POLLINATION MECHANISM OF CORONILLA

In coronilla (*Coronilla varia*), the upper edges as well as the lower edges of the keel petals are fused, but a small opening remains at the tip of the keel.

When the anthers open, the pollen accumulates only in the tip of the keel because the filaments block the way to the rest of the keel. When a bee depresses the keel, the stamens remain in position and so push the pollen out of the opening of the keel. Some of this pollen touches the bee's body and clings to it. Depressing the keel also forces the stigma out of the opening, and it may receive pollen that the bee brings from another flower. When the bee leaves, the keel returns to its original position. This pollination mechanism can be worked again by bees visiting the flower later. Lupines and bird's-foot trefoil (*Lotus corniculatus*) have similar pollination stories.

The vetches (*Vicia* spp.) have a pollination story similar to that of bird's-foot trefoil, but in these flowers, the pollen becomes caught on the *stylar brush,* a cluster of hairs on the style of the pistil. This happens shortly before the flower is ready for pollination. When a bee presses down on the keel, the stigma and stylar brush are exposed, and some of the pollen on the stylar brush is deposited on the bee. The stigma may receive pollen from the bee at the same time. This mechanism, too, can be worked several times. The everlasting pea (*Lathyrus latifolius*) has a pollination story similar to that of the vetches.

In alfalfa the pollination mechanism is an explosive one. The stamens and pistil are held in the keel under tension. When a bee depresses the keel, the

STAMENS AND PISTIL

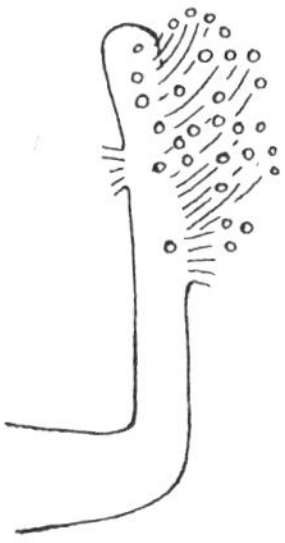

STYLAR BRUSH
WITH POLLEN

VETCH

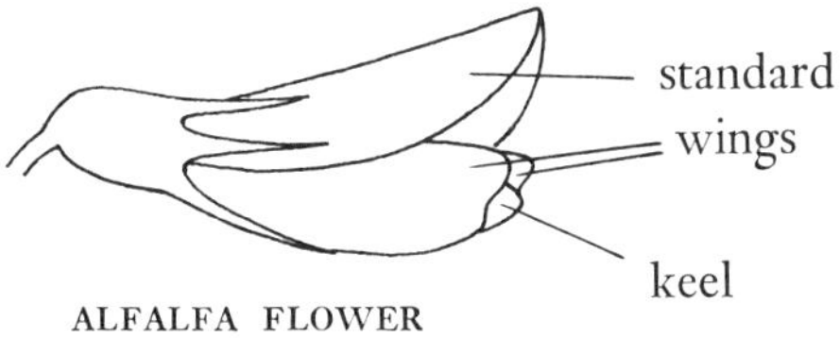

ALFALFA FLOWER

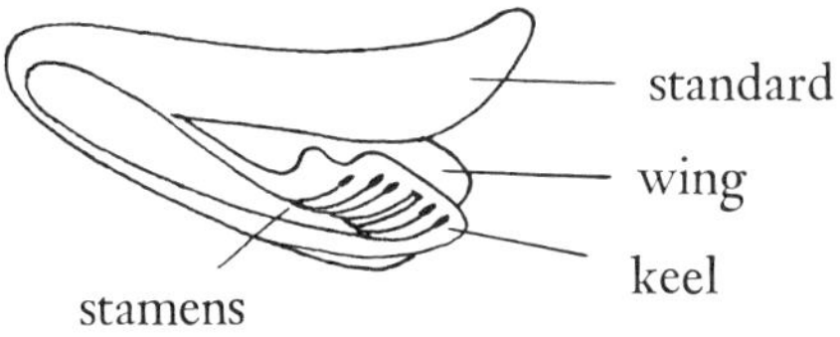

SECTION OF FLOWER

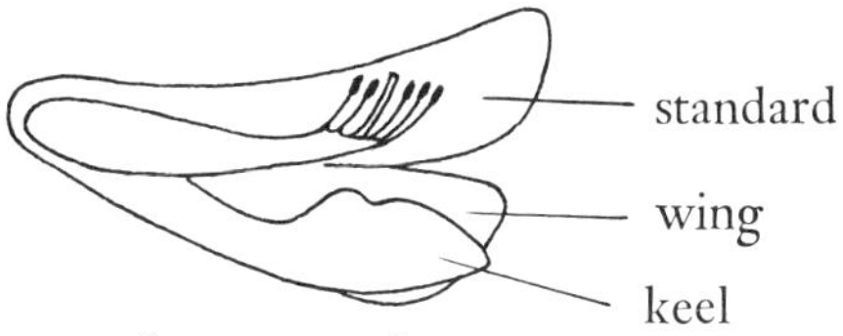

STAMENS "EXPLODED"

POLLINATION MECHANISM OF ALFALFA

tension is released, and the stamens and pistil snap upward explosively, dusting the lower surface of the bee with pollen. At the same time the stigma receives pollen previously dusted onto the bee by another flower. The explosive pollination mechanism works so quickly that bees may become momentarily pinned against the standard by the stamens and pistil. This happens only to bees seeking nectar, for they must insert their heads farther into the flowers than pollen-seeking bees do. Becoming trapped by the flower, even for only a moment, seems to be annoying to bees, for after having had this experience a few times, nectar-seeking bees soon learn to bite holes into the side of the keel and to steal the nectar through the holes without becoming caught but also without pollinating the flowers. The explosive pollination mechanism can be operated only once, for the keel does not return to its original position and bees rarely visit an exploded flower.

Scotch broom (*Cytissus scoparius*) has a similar but somewhat more spectacular explosive mechanism. The flowers have ten stamens, five long ones about the same length as the pistil and five short ones. When a bee depresses the keel, the five short stamens spring upward and touch the lower surface of the bee's abdomen. At the same time the long stamens and the pistil curve upward and touch the upper surface of the bee's abdomen. Small clouds of pollen grains released from the anthers surround the bee and dust it on all surfaces.

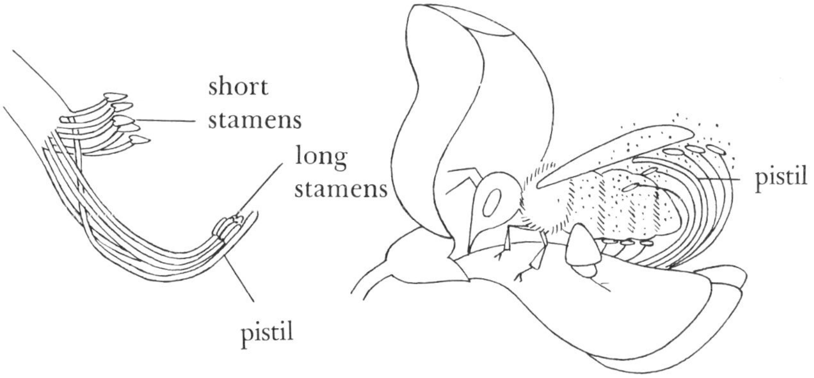

POLLINATION MECHANISM OF
SCOTCH BROOM

Among bird-pollinated leguminous plants are the coral trees (*Erythrina* spp.). The cockscomb coral (*Erythrina crista-galli*) from Brazil is the species most commonly cultivated in southern United States. Its

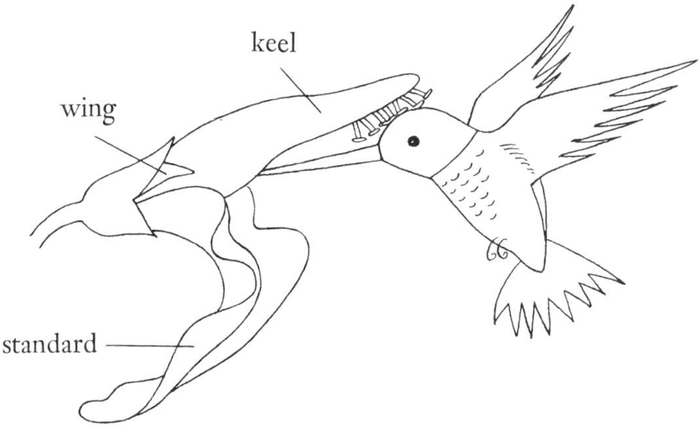

POLLINATION OF COCKSCOMB

red flowers are visited by hummingbirds. These birds hover as they feed; that is, they stay in one position in the air without flying forward—somewhat like a helicopter hovering in one position. Flowers of the cockscomb coral tree hang upside down, and the keel, which is rather stiff, is the uppermost part of the flower. Because the anthers and stigma extend downwards from the keel, they are fully exposed to pollinating birds, and the pollination mechanism does not require the keel to be moved as in so many other flowers of the pea subfamily. The standard hangs downward, and the wings are very small and inconspicuous. Nectar is produced in abundance in the keel, and when hummingbirds insert their beaks into the keel to obtain the nectar, their heads bump against the anthers and stigma, thus bringing about pollination.

Perhaps the simplest pollination stories are those of plants like garden peas or sweet peas, which are regularly *self-pollinated.* In this case, pistils are pollinated with pollen from the stamens of the same flower. The stamens and the pistil remain confined within the keel, and some of the pollen touches the stigma. Other plants that usually are self-pollinated are kidney beans, lima beans, soybeans, and peanuts, although bees may occasionally cross-pollinate them as well.

If you have bean plants or pea plants in bloom in a vegetable garden, you should be able to see the ripening of pistils into mature pods. Open one of

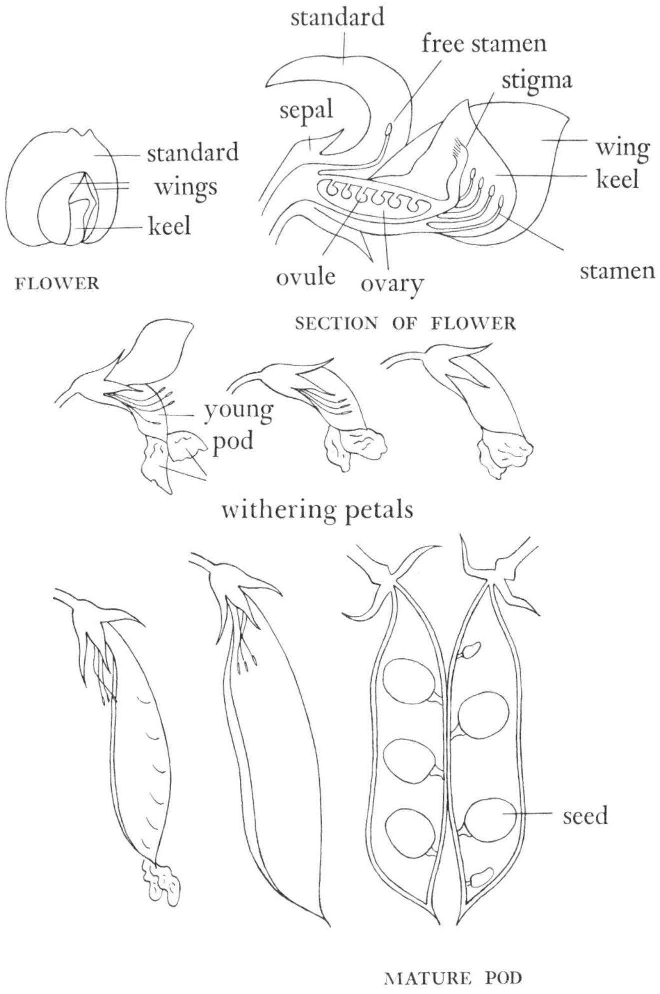

DEVELOPMENT OF A PEA POD
FROM AN OVARY

the freshest-looking flowers and remove the petals and stamens to expose the pistil. Even though the pistil is small at this stage, its ovary resembles a miniature bean or pea pod. If you look carefully, you probably can just barely see the tiny ovules; use a magnifying glass if necessary. Look also for flowers with wilted petals; many of these flowers have been pollinated recently, and their ovaries may be just a little larger than those with fresh petals. Now look for flowers from which the petals are falling; the ovaries will look a little more like the mature pods. You probably can find pods that are still older and have larger ovules that are really young seeds by now; the petals probably have all fallen, but the sepals still remain. Often you can find the dry, brittle remains of stamens at the base of a ripe pod.

The seeds of many members of the pea subfamily are spherical (like peas) or kidney-shaped (like beans). Each seed contains one *embryo,* or young plant, which fills nearly the entire seed. Only a thin seed coat covers the embryo. The seed coat may be soft, as in green garden peas; hard, as in the black locust (*Robinia pseudoacacia*); or papery, as in peanuts.

While still developing, the seed is attached to the pod by a stalk. When the mature seed breaks from the stalk, a scar called the *hilum* is left on the seed. The hilum is usually easy to see on most bean and pea seeds. It is the "eye" of black-eyed peas (*Vigna sinensis*); the black part, however, is a pair of black lines that nearly surround the hilum.

The root tip of a bean or pea embryo lies near the hilum just below the seed coat. Sometimes the root tip raises the seed coat slightly so that a small ridge seems to point toward the hilum. It is quite obvious in fresh peas, and you can see it in canned peas if you look closely. Pea seeds obtained from a seed store usually are too dry and wrinkled for these details to be clear, but if you soak the seeds in water for a few hours you should have no difficulty seeing them. The root tip especially is quite obvious in well-soaked pea seeds.

The embryo consists of a short stalk with the root tip at one end, a tiny stem tip at the other end, and

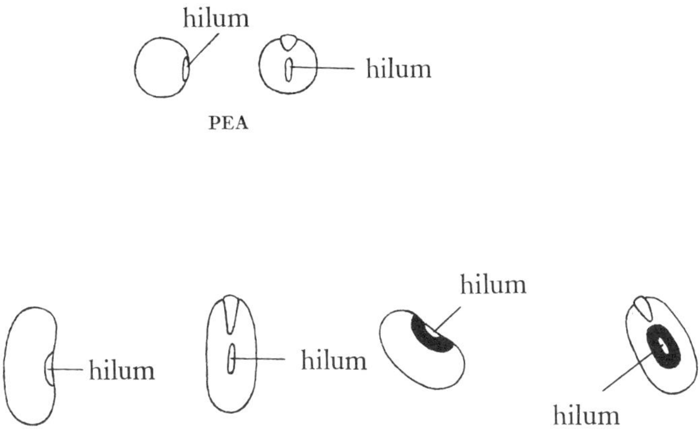

GREEN BEAN BLACK-EYED PEA

SEEDS

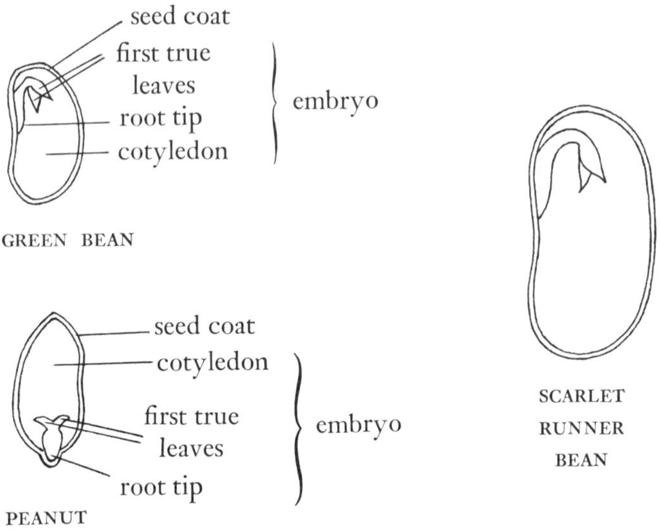

SECTIONS OF SOME SEEDS

two enormous, thick leaves called *cotyledons,* or seed leaves. When you open a bean or pea seed, one of the first things you notice is that its contents can easily be divided into two halves; each of these halves is a cotyledon. If you pull them apart, you will find the remainder of the embryo attached to one of them. The root tip is easy enough to see, but the stem tip may be a little more difficult to find, for small, flat leaves, the first true leaves of the young plant, surround it.

If you plant an intact seed, the young root will grow; from it will come branch roots, and these may have branch roots as well. Through its roots the plant will obtain water and minerals from the soil. The young stem tip will grow above the ground. It, too, may branch, and it will produce leaves, and later flowers, fruits, and seeds. In sunshine, the leaves will manufacture food, which supplies the plant with the energy it needs to grow and to reproduce. The very first growth of the young embryo, while it is still within the seed, however, depends on the food stored in its cotyledons. The embryo plant received this food from its parent plant after pollination and while the ovary was growing into a mature pod.

Chapter 4

Some Members of the Pea Subfamily

THE SOYBEAN (*Glycine max*), so little known until recently in the United States and Europe, has been one of the main foods of the people of eastern Asia since before recorded history. Its protein content is higher than that of other leguminous crops and of meat and other animal products as well. No one knows when soybeans were first cultivated, but they well may be the world's oldest food crop. Over the centuries, Oriental peoples learned to prepare them in many different ways.

You can eat soybeans as a fresh vegetable, either raw or cooked. Soybean seeds are about the size and shape of peanut seeds or perhaps a little smaller; the seed coat is not papery like that of peanuts, though. Eaten raw, soybeans have a slightly peanutty flavor.

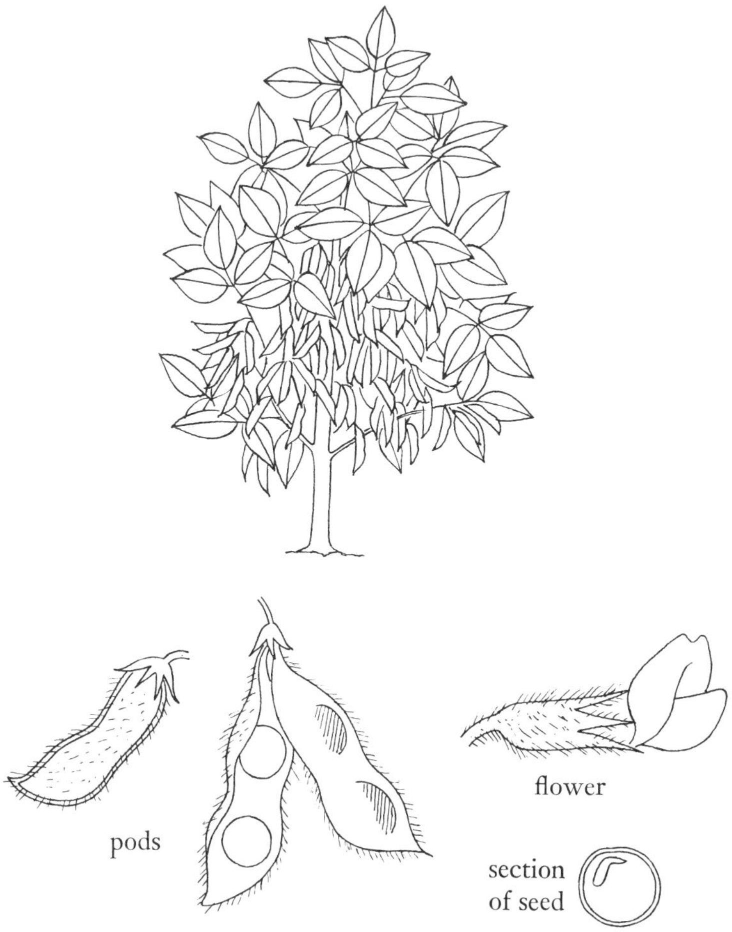

SOYBEAN

Fresh or dried soybeans can be cooked as a vegetable, too. Cooked soybeans have little flavor, and people usually mix them with other foods to make a tastier dish. Meatless casseroles made with soybeans can be quite nutritious.

Raw soybeans can be sprouted for use as bean sprouts. When seeds sprout, their vitamin C content increases greatly, and this makes them even more nutritious than the dry seeds. The small bean sprouts so common in Oriental cuisine are a different species, *Phaseolus aureus,* the mung bean, but both can be sprouted the same way. The main thing to remember in sprouting beans is to keep the seeds moist but not soaked in water. Now that sprouting beans in the home is becoming more fashionable in the United States, you can purchase a sprouter in most diet food stores. But you need not go that expense; you can easily make your own sprouter out of two wide-mouthed jars of slightly different size. The neck of one jar should fit into the neck of the other as shown in the illustration. Soak the beans in water for a few hours in the jar with the smaller opening. The beans will swell to at least twice their size, and later, when roots form, they will need even more room. So it would be best to fill the jar not more than about one-sixth full of dry beans. This will allow space for air to circulate later. When the beans have soaked for several hours, cut a piece of cheesecloth large enough to cover the opening of the jar and secure it there with string. Then drain off the water by letting it run out through the cheesecloth. Put some fresh water in the jar with the larger opening—about an inch deep—the exact depth is not important. Now invert the jar with the beans into the mouth of the jar with water. Place them in a dark cabinet. About three times a day (four or five times if it is summer and the cabinet is hot)

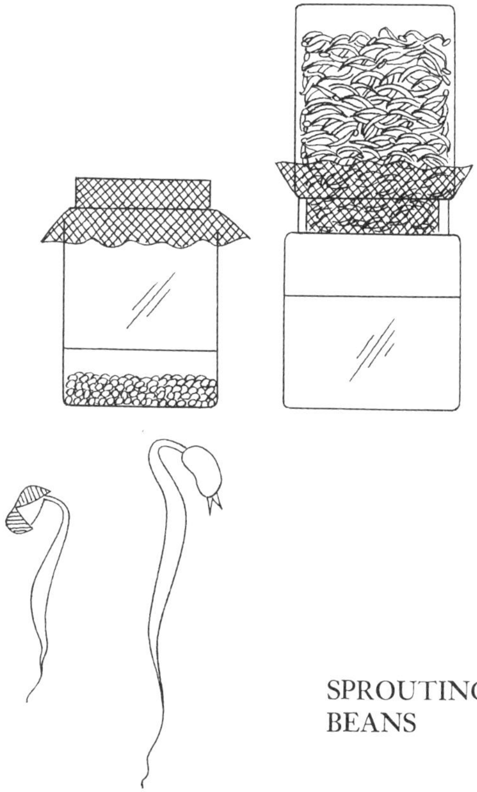

SPROUTING BEANS

rinse the seeds with lukewarm water and replace them in the cabinet. This arrangement keeps the seeds moist, but air can still circulate among them. (If the beans are constantly wet they will soon mold and be useless. Always discard any seeds or sprouts that appear to be moldy.)

In one or two days you will see the white root tips emerging from the seeds. In another day or two the roots will be about an inch long. The sprouts can be eaten then, or you may prefer to let them grow just a little longer. The entire process may require only two or three days in hot weather and perhaps five or six in winter. Soybean sprouts make a wonderful snack just as they are; like the raw soybeans, they have a slightly peanutty flavor, but they are fresh and crisp. You can add raw sprouts to a tossed salad. Sautéed in a little butter, they can be added to many hot dishes, such as casseroles, many hot vegetables, or scrambled eggs (for scrambled eggs you may prefer the smaller mung bean sprouts). Sprouts can be cooked by themselves as a vegetable, too.

You probably won't want to eat all the sprouts as soon as they are ready, but you can keep them in the refrigerator for several days just as you would keep fresh vegetables purchased in the supermarket.

Always use seeds intended for sprouting, such as those sold in health food stores. Seeds sold in garden shops are intended for planting and may have been treated with poisonous chemicals to prevent their becoming moldy or being eaten by insects.

Soy flour ground from soybeans is much richer in protein than the flour from cereal grains. It can be used as a substitute for a portion—usually from 5 to 20 percent—of wheat flour in many recipes. In Israel, soy flour is added to bread flour, which by law must contain 10 percent soy flour. Bread, cookies, and cakes are made more nutritious this way, and the need for meat in the diet is thereby reduced.

Today food biochemists are producing new meat substitutes from soybean protein. They extract protein from the beans and spin it into a fiber, which they combine with suitable flavorings and coloring agents and fashion into a form and texture that resembles bacon, sausages, ham, or chicken. These products cost somewhat less than real meat. Their fat and cholesterol content is considerably lower than that of meat, and this is an added attraction to persons who worry about getting too much of these substances in their diet.

Some other soybean products are soy sauce (made by fermenting cooked soybeans and roasted wheat for several months or even years), soy milk (prepared by cooking ground soybeans and then straining the mixture to remove the solid matter), and bean curd. Bean curd, made from soy milk, is also called tofu and soy cheese, but its flavor resembles no flavor that Americans or Europeans expect from their cheeses. Soybean oil is used to make margarine, salad oil, paints, varnish, linoleum, soap, printing ink, candles, and other products. Soy butter is made from soybean oil and soy flour or ground soybeans.

Ancient farmers of the Middle East and the Mediterranean area cultivated the broad bean (*Vicia faba,* also known by the common names of horse bean, English bean, and Windsor bean) long before written history. Its seeds have been found preserved in Bronze Age excavations in Switzerland, and it was cultivated later by the ancient Hebrews, Egyp-

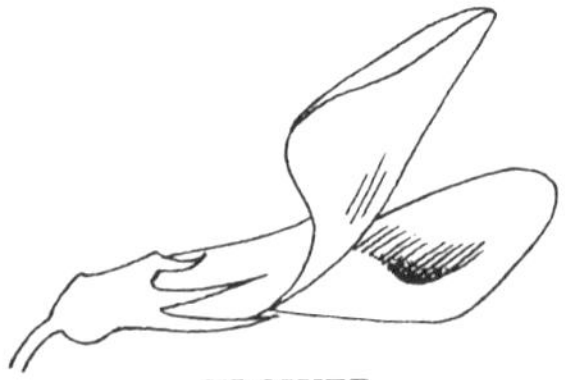

FLOWER

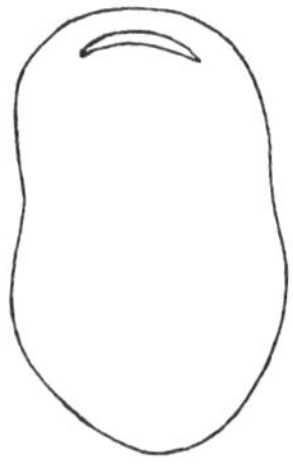

SEED

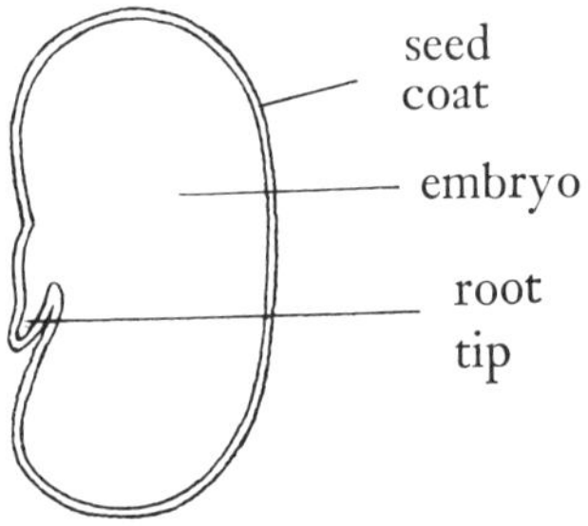

SECTION
OF SEED

BROAD BEAN

tians (who sometimes placed the beans in mummy coffins for use by the dead in the spiritual world), Greeks, and Romans. Broad beans still are a staple food in southern Europe and northern Africa, but they have never found much favor in America except among persons of Mediterranean descent.

The seeds range in color from white to brown, green, purplish, and black. In ancient times they were used in voting: a white seed indicated a pro vote, a black one a con vote.

Another leguminous crop of ancient times in the Middle East was the lentil (*Lens esculenta* or *Lens culinaris*). It is still an important crop from the Mediterranean area eastward to India and Pakistan. Lentil plants do well on poor ground, and so farmers raise them where other crops will not grow. The

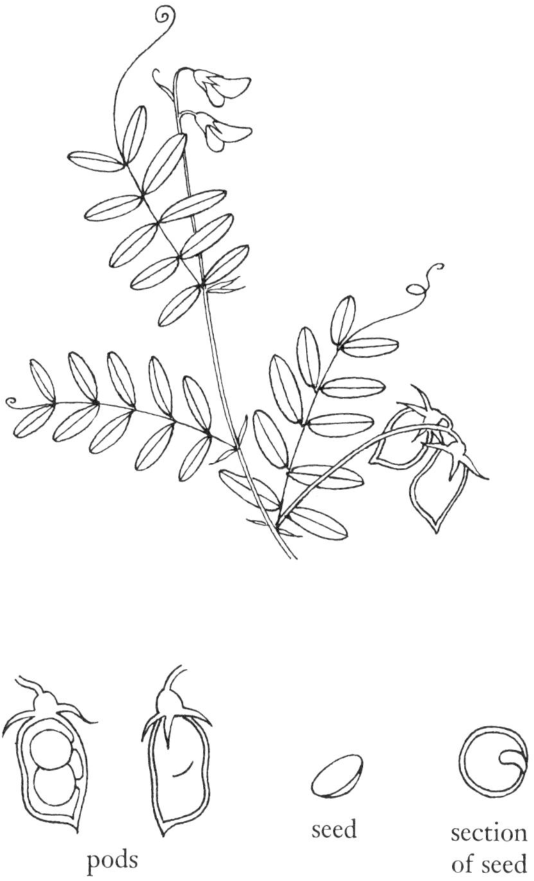

LENTIL

seeds are used in soups and to make porridge. The Old Testament mentions lentils several times. Genesis 25: 29–34 records the story of the hungry Esau who sold his birthright to his twin brother Jacob for bread and a pottage (thick soup) of lentils.

Lentils must have been well known in Europe when the first small magnifying glasses were made, for these little pieces of ground glass were named lenses for the generic name of the plant whose seeds they resembled in size and shape.

You can sprout lentils the same way you sprout soybeans and mung beans.

The common garden pea (*Pisum sativum*) was another ancient crop of the Mediterranean area. Its seeds, too, have been found in dwellings dating back to the Bronze Age. Throughout most of history, people gathered only mature, dry pea seeds for food. Then, in the sixteenth century, they began to eat the immature seeds called green peas or sweet peas (but these are not the same as the ornamental sweet pea, *Lathyrus odoratus*). Near the end of the seventeenth century, when Louis XIV reigned in France, eating green peas was so fashionable that even people who had dined well at court would hurry home to have a snack of green peas before going to bed. Today we accept green peas as a common vegetable and use the dry peas mostly for split pea soup. The name sugar pea is usually applied to varieties of peas with edible pods.

The garden pea played a special role in the study

of genetics, the science of heredity. During the nineteenth century, Gregor Mendel, an Austrian monk, performed experiments with garden peas in order to determine how several of their characteristics are inherited. Although scientists ignored his work during his lifetime, his experiments later became the basis of our modern science of genetics. Today many biology students are assigned homework problems dealing with inherited traits of pea plants—such as the height of the plants, the color of their flowers or seeds, and whether the seeds are smooth or wrinkled.

If someone mentions black-eyed peas (*Vigna sinensis*), you probably automatically think of the southern United States where these peas are so often on

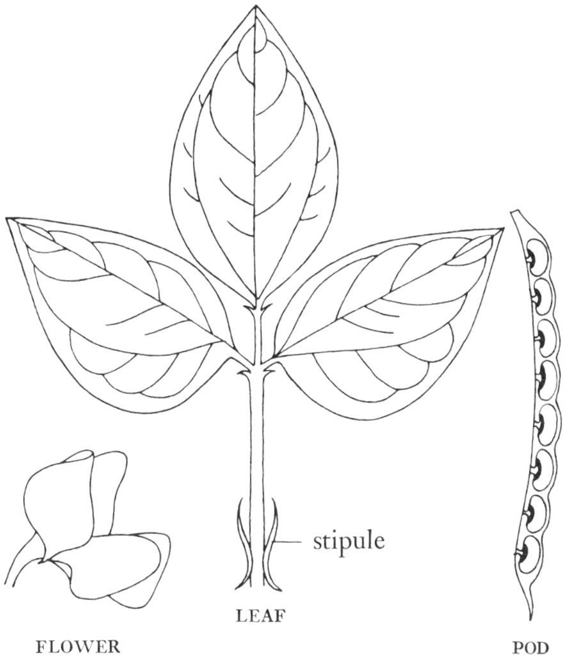

BLACK-EYED PEA

the menu. Although they seem to be a part of the South, they are not native to the United States at all. Their country of origin is unknown and is variously believed to be somewhere in tropical Asia or central Africa. Another name for them is cowpea.

A close relative of black-eyed peas is the yard-long bean (*Vigna sesquipedalis*). Though edible it is often grown merely as a curiosity because of its long pods. However they rarely reach the length suggested by their common name. The specific epithet *sesquipedalis* means one and a half feet and is perhaps a better description. Yard-long beans are also called asparagus beans.

The American Indians cultivated several varieties of *Phaseolus vulgaris*. Some of these varieties are known by the common names of kidney bean, pinto bean, and great northern bean. These beans are picked in their dry, mature state; the seeds are saved, and the pods are discarded. *Phaseolus vulgaris* also includes green, string, or snap beans and a yellow variety called wax or yellow bean. We eat the immature pods along with the immature seeds inside. The pole varieties of *Phaseolus vulgaris* are vines that require supports on which they can twine. Bush or dwarf varieties are shorter and need no support. *Phaseolus vulgaris* has been eaten by American Indians for at least 8,000 years; seeds were discovered in Peruvian graves dating from 6,000 B.C.

Lima beans (*Phaseolus lunatus*) are also native to the Western Hemisphere. They have been found in

FLOWER

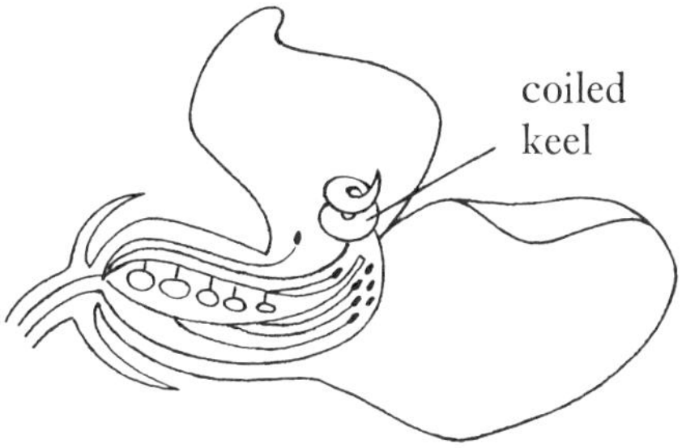

SECTION OF FLOWER

GREEN BEAN

ancient excavations in Peru, and they probably were cultivated there about 4,000 B.C.

Nearly all species of *Phaseolus* are native to the American tropics, but a few are Asian. *Phaseolus*

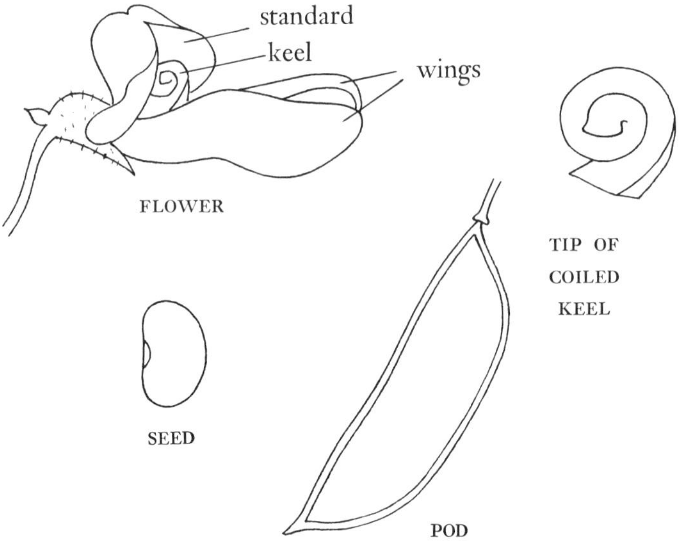

LIMA BEAN

aureus, the mung bean or golden gram bean, which probably comes from India, ·provides the bean sprouts so much a part of Oriental cookery.

The peanut (*Arachis hypogaea*) is one of the few plants with seeds and fruits that develop underground. After pollination, the yellow petals of the flower whither quickly and fall. The stem below the ovary elongates and bends downward until it forces the ovary into the ground. If, for any reason, the ovary does not enter the soil, no fruit or seeds form. The peanut (also called goober and groundnut) is believed to be native to Brazil, but now it is grown in the southern United States and in the tropical

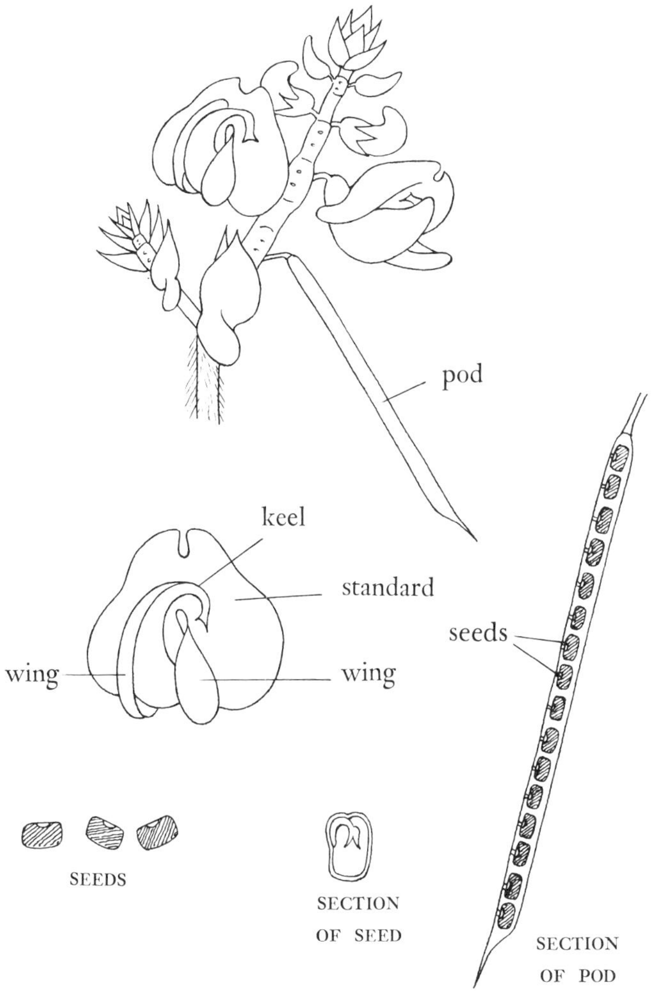

MUNG BEAN

regions of Africa and India. The seeds may be eaten raw, but more than half of the peanuts raised in the United States are used for peanut butter, and roasted

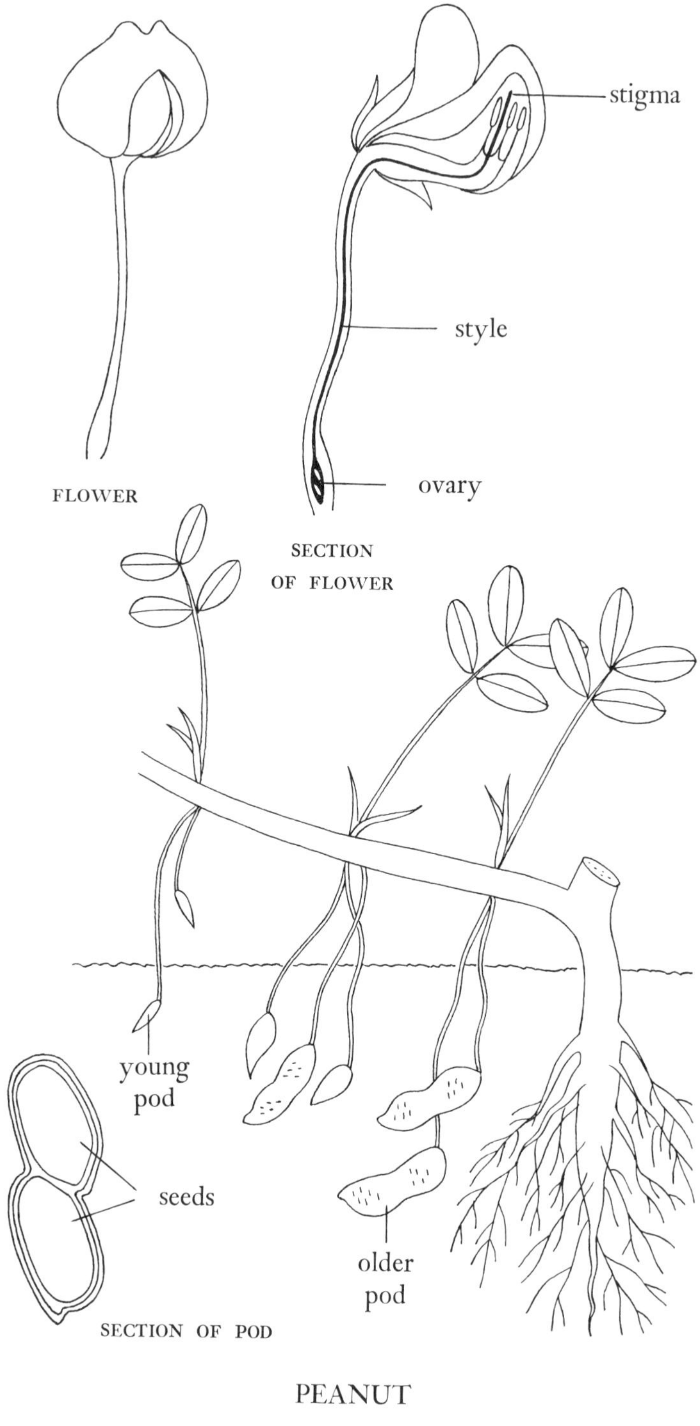
stigma
style
ovary
FLOWER
SECTION
OF FLOWER
young
pod
seeds
older
pod
SECTION OF POD

PEANUT

peanuts seem to be an all-American snack. Peanut oil, like soybean oil, is used in margarine, salad oil, soaps, plastics, and lubricants. Farmers use the entire plants as forage. They sometimes turn pigs loose in a field of peanuts, where pigs eat not only the stems and the leaves but dig up the pods as well—a great convenience to the farmer who does not have to harvest the crop.

Most of the leguminous plants raised as food for human beings make excellent forage for livestock as well, and farm animals frequently are fed the parts left after the seeds or fruits are harvested. A few species, however, are raised primarily as forage. Alfalfa or lucerne (*Medicago sativa*), called the "queen of forages," is considered the best protein source among forage plants. Its taproot grows deep into the soil where it can absorb water from great depths; for this reason alfalfa is resistant to drought and can be grown in a wide variety of climates.

Alfalfa is believed to be native to Iran, where seeds have been found in archeological deposits dating back to about 4,000 B.C. The ancient Persians raised it for their chariot horses about 500 B.C. When Spaniards explored Central and South America, they brought alfalfa with them to feed their horses.

Some persons eat the leaves and flowers of alfalfa and alfalfa sprouts, but most people find them rather tasteless. That problem can be solved by mixing them with other more tasty greens in a tossed salad or sprinkling the dried and powdered leaves into soups and stews.

ALFALFA

Second in value to alfalfa as forage plants are the clovers (*Trifolium* spp.). *Trifolium pratense* is the red clover widely cultivated in North America. Its purplish-red heads are a common summer sight in fields and along roadsides, where it grows wild. Other clovers grown for forage are alsike clover (*Trifolium hybridum*) with white or pink flowers, crimson clover (*Trifolium incarnatum*) with bright red flowers, and white clover (*Trifolium repens*), the small clover with white or very pale pink flowers that grows as a wild flower or weed in lawns.

Like alfalfa, clovers occasionally are eaten by human beings and in much the same way. In times of famine, people may crumble the dried leaves and flowers and add them to flour in order to extend the

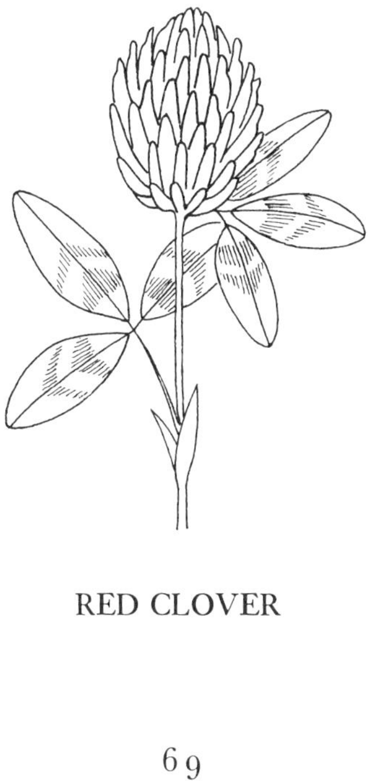

RED CLOVER

flour. The seeds, too, have been eaten in times of emergency.

Bees make honey from the nectar of many flowers, but clover honey has always been a favorite. Honey was the main sweetening agent in Europe before people learned to extract sugar from sugar cane and sugar beets.

White sweet clover (*Melilotus alba*) and yellow sweet clover (*Melilotus officinalis*) contain a substance called coumarin that is responsible for the

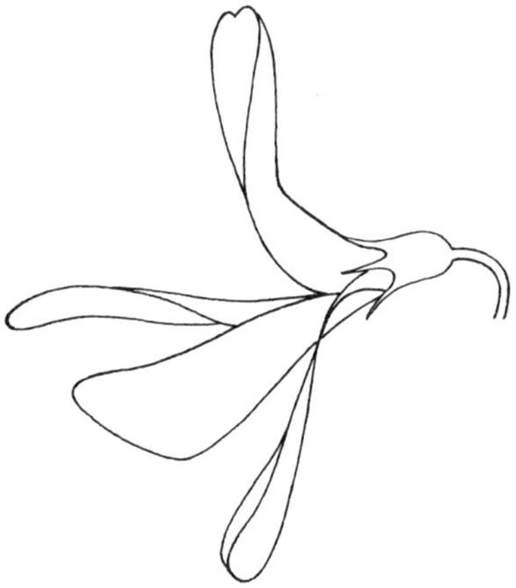

FLOWER

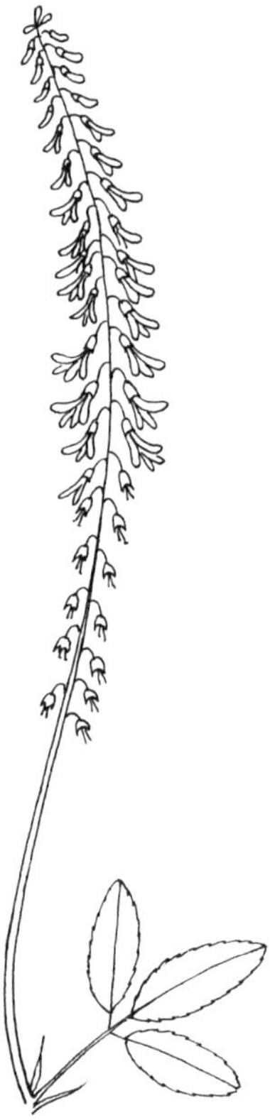

YELLOW SWEET CLOVER

delightful scent of new-mown hay. The fragrant flowerheads can be dried and used in sachets, and the dried leaves are sometimes used as a substitute for vanilla in cookies and puddings.

A number of wild flowers in the pea subfamily have edible parts. The ground nut (*Apios tuberosa*), also called wild bean, potato bean, and hopniss, is credited with enabling the Pilgrims to survive their first few winters in New England. In July and August the plants bear fragrant, chocolate-brown flowers;

GROUNDNUT

these are followed by pods containing edible seeds that can be boiled like an ordinary vegetable. Of more importance are the edible tubers arranged on an underground stem like beads in a necklace. Though frequently compared with small potatoes, they taste a little more like turnips. They can be eaten raw, but persons who have done so complain that the juice forms a rubbery scum on the teeth and lips. They are much better boiled or roasted. The ground nut grows in moist woods throughout much of the eastern half of the United States.

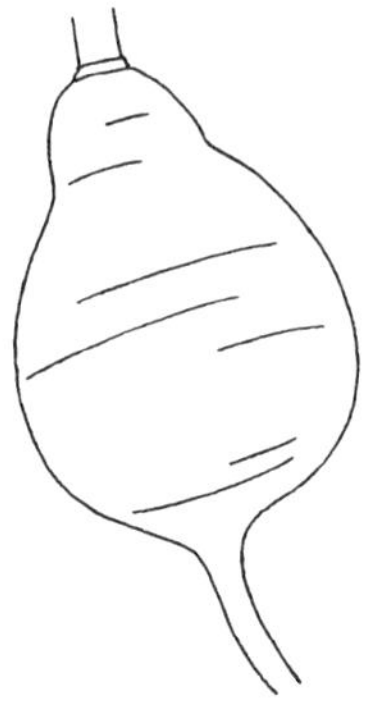

PRAIRIE-TURNIP ROOT

Farther west, in the prairies and plains, the prairie-turnip (*Psoralea esculenta*), also called bread-root, prairie apple, and pomme blanche, has a starchy root that may be prepared like potatoes; raw, it, too, has a turniplike flavor. Still farther west, in the deserts, grows the beaver-dam breadroot (*Psoralea cas-*

torea); it, too, has a starchy root. The Indians and early settlers used both of these plants for food.

Wild licorice (*Glycyrrhiza lepidota*), which grows in moist prairies, has an edible root that the Indians and early settlers roasted or chewed raw. Licorice used for flavoring candies and tobacco comes from the dried roots and rhizomes (underground stems) of *Glycyrrhiza glabra*, which is cultivated in Europe and Asia. Licorice has a soothing effect on sore throats and is used to flavor cough drops.

The pea subfamily contains few plants used as spices, but fenugreek (*Trigonella foenum-graecum*), of southern Europe and Asia, has pungent seeds used in curry powders, chutney, and imitation vanilla extract, as well as some perfumes. The plants, which are cultivated primarily in India and North Africa, are also eaten as a vegetable or fed to livestock.

Most of the trees in the pea subfamily are tropical, and only a few are native in the United States.

Black locust (*Robinia pseudoacacia*), also called white locust, yellow locust, and false acacia, grows in the eastern part of the United States. When it was introduced into Europe, botanists there believed it to be an acacia, but acacias are members of the subfamily Mimosoideae. A reminder of the error lingers on in one of its common names and in its specific epithet; a perfume once made from the fragrant white flowers is also called acacia, but acacia perfume now is made synthetically.

Because black locust wood does not rot when in

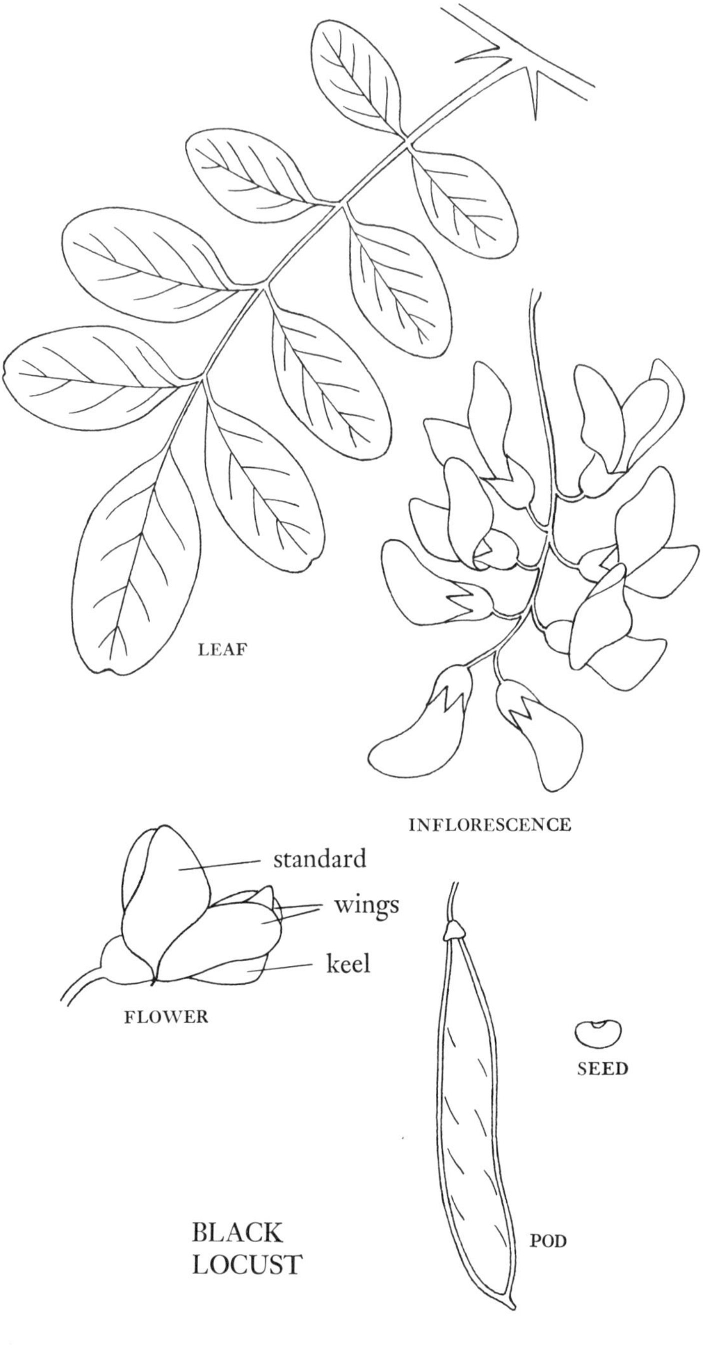

BLACK
LOCUST

contact with the soil, it has been much in demand for fence posts, railroad ties, and vine stakes in vineyards. It was once used for wooden nails because it did not swell in damp weather or shrink in dry weather as most other woods do. The insulator pins on the cross arms of telephone poles and power line poles are made of black locust wood.

Some other locusts of this genus are clammy locust (*Robinia viscosa*) and bristly locust or rose-acacia (*Robinia hispida*), both of eastern United States, and New Mexican locust (*Robinia neo-mexicana*) of southwestern United States. These species have pink flowers.

The deserts of southwestern United States, which are dull and drab for much of the year, display brilliant colors in spring and early summer when desert plants bloom. Usually we think of these plants as being cactuses. However, a few are leguminous plants, and people who have seen them in bloom describe them as being covered with glorious and spectacular colors.

The desert ironwood (*Olneya tesota*) has deep blue to rose-purple flowers, and bright blue flowers cover the branches of smokethorn or indigo bush (*Dalea spinosa*). Both of these plants are small trees. The desert ironwood receives its common name from the fact that the wood of the dead trees is very hard. Indians used it to make points for their arrows. The wood is unusual in another way: it is so heavy that it sinks in water. Because it burns slowly and

DESERT IRONWOOD

produces a pleasant odor, the wood is often used for firewood.

Before the twentieth century, dyes were obtained from plant and animal sources rather than being manufactured in chemical factories as many of them are now. Fast dyes (dyes that do not fade when washed or exposed to sunlight) were in great demand, and the "king of dyestuffs" was indigo, a blue dye obtained from the indigo plant (*Indigofera tinctoria*) of the East Indies. The dye was extracted from the leaves of the plant and packaged as dry cakes to be shipped to Europe. Although indigo had been used to dye cloth for at least 4,000 years, so little did Europeans understand about its origin that in 1705 Britain granted a patent for mining indigo. Today, a cheaper, synthetic dye replaces natural indigo.

Several species of wild indigo (*Baptisia* spp.) grow in the United States, and early settlers obtained a blue dye from them, although it was inferior to that of *Indigofera*.

In spite of the fact that so many plants in the pea subfamily produce nutritious food, a few of them are poisonous. Lupines, sweet peas, wisteria, and the locusts (*Robinia* spp.) all have poisonous parts. The rosary pea (*Abrus precatorius*), also called jequirity bean, precatory bean, and crab's-eye, has beautiful red seeds, each with a black spot at one end. The seeds, which have been used in costume jewelry and in rosaries, contain a powerful poison and should never be eaten.

The locoweeds, which belong to the genera *Astragalus* and *Oxytropis*, are weeds of southwestern and south central United States. There are many species of them, and some can be poisonous to livestock. Most livestock prefer to eat other plants, but some, horses especially, can acquire a taste for them. The plants contain a narcotic substance called locoine, which makes the animals act "loco" or "crazy" if they eat enough of the plants. Horses, for instance, will take great leaps just to cross over some pebbles. Paralysis and death may occur.

In some western states the soil contains high concentrations of the element selenium, and species of *Astragalus* can accumulate the selenium in their tissues. Animals grazing on the plants become ill or die from selenium poisoning.

In several tropical areas of the world, native peoples learned to use different species of "fish poison" plants of the pea subfamily to make fishing a little easier. Fishermen remove branches from the plants, beat them to damage the tissues, and then throw them into a stream. The sap, which contains the poison, seeps into the water and stuns the fish. The fishermen merely wade into the water and pick up the fish. The poison in no way spoils the fish for human consumption.

The poison in these "fish poison" plants is rotenone, and it has a use in addition to stunning fish. It is a natural insecticide, and before the development of synthetic insecticides, it was one of the three major substances used to kill insects. Farmers kill the European corn borer, the pea aphid, the woolly apple aphid, and the Mexican bean beetle with rotenone. It also kills fleas, ticks, and lice on livestock. Houseflies, cockroaches, and mosquitoes are susceptible to it; but honeybees, fortunately, are relatively resistant. Though partially replaced by synthetic insecticides, rotenone has an advantage over them because it is biodegradable—that is, it is destroyed quickly by bacteria or other microorganisms, and so does not continue to kill insects or other animals when it is no longer needed.

Two of the major commercial sources of rotenone are the roots of species of *Derris* in the Far East and of *Lonchocarpus* in South America. The coral trees

(*Erythrina* spp.), mentioned later in this chapter for their beauty, are used by South American Indians to poison rats as well as to stupefy fish.

The kudzu vine (*Pueraria lobata*) is a woody vine from China and Japan, where it has been grown for forage as well as for ornamental purposes. The

TREES COVERED BY KUDZU VINES

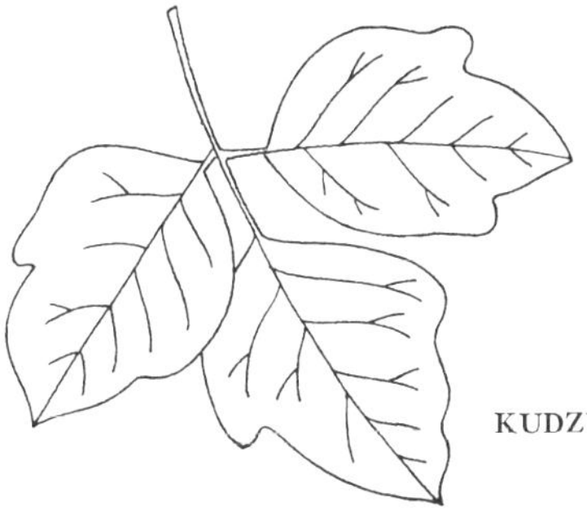

KUDZU LEAF

Japanese people extract starch from kudzu roots. Because the plant grows rapidly and enriches the soil by nitrogen fixation, it was brought to the southeastern United States to prevent erosion and to improve soils that had been impoverished by many years of raising cotton and tobacco crops. Unfortunately, the kudzu adapted to its new home even better than had been expected. It became a pest throughout much of southeastern United States, where it climbs over other plants, shading them from the sun and thus killing them. In some places, kudzu vines cover entire forests.

Some of the most beautiful ornamental plants and wild flowers are in the pea subfamily. Sweet

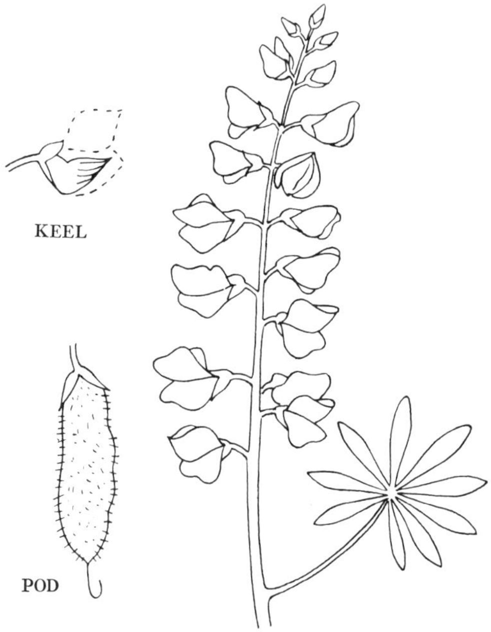

LUPINE

peas (*Lathyrus odoratus*) come in several colors and are enjoyed in the garden and as cut flowers. Several species of lupine (*Lupinus* spp.) are grown in gardens, and some are wild flowers. The Texas bluebonnets, which cover plains and hillsides with a blanket of blue when they bloom, are *Lupinus subcarnosus* and *Lupinus texensis.* The name *Lupinus* means "wolf flower" and was given to these plants when it was believed that they devoured the minerals in the soil and left it poorer than before they grew there. Actually, like most other leguminous plants, they enrich the soil with nitrogen.

Leadplant (*Amorpha canescens*) neither absorbs lead nor serves as an indicator of lead deposits in the

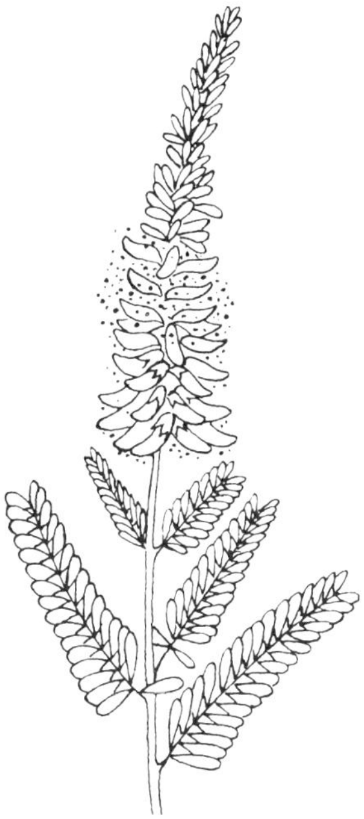

LEADPLANT

PURPLE PRAIRIE CLOVER

soil. A covering of very fine hairs gives it a grayish, leaden hue—and its name. Leadplant and the purple prairie clover (*Petalostemum purpureum*) rarely grow in disturbed soils; their presence usually indicates that the land they grow on is virgin prairie. The flowers of leadplant are described on page 34 and illustrated on page 35.

The wisterias are woody vines. Some are native in southeastern United States, but two of the most commonly cultivated species are the Japanese wisteria (*Wisteria floribunda*) and the Chinese wisteria (*Wisteria sinensis*). An arbor arching over a pathway allows their lovely drooping racemes to be displayed to advantage.

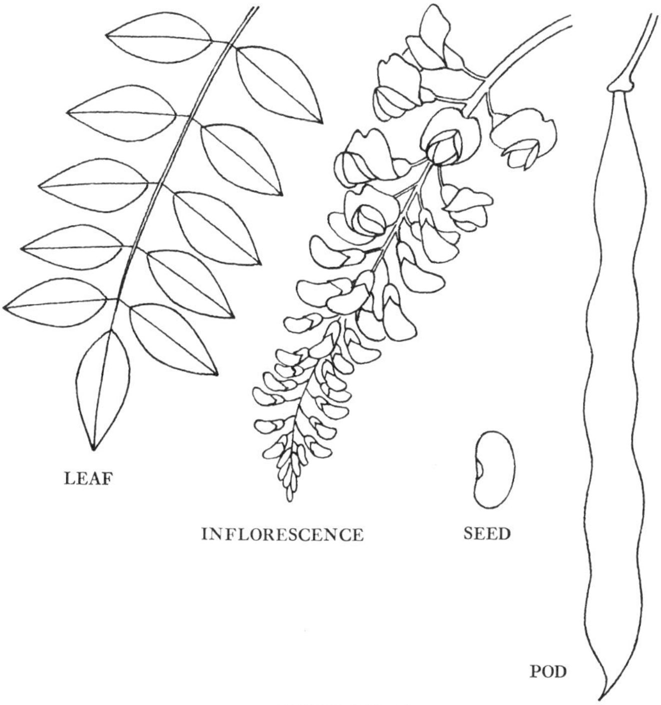

WISTERIA

Some of the most spectacularly beautiful leguminous plants are natives of tropical and subtropical areas of the world. People living in colder climates rarely have an opportunity to enjoy them. Some of the coral trees or immortelles (*Erythrina* spp.) have blood-red flowers. Their pollination by hummingbirds is described in the preceding chapter. *Erythrina herbacea* is an herb with similar flowers; it grows in sandy soil along the southeastern coasts of the United States from North Carolina to Texas.

It seems a shame that the jade vine (*Strongylodon macrobotrys*) is confined to the tropics. This woody vine from the Philippine Islands produces hanging racemes that may be five feet long and crowded with flowers, the most unusual feature of which is their brilliant blue-green color—rare among flowers.

Chapter 5

The Senna Subfamily

(*Subfamily Caesalpinioideae*)

IF YOU WERE to see one of the members of the senna subfamily in fruit, you would recognize it instantly as a member of the family Leguminosae, for the fruits are legumes or pods similar to those of the pea subfamily. The leaves, too, have the general appearance of the leaves of beans and peas; a few are simple, but most are pinnately compound. If you examined the flowers, however, you probably would not see the resemblance so quickly, for with a few exceptions, the flowers are not butterflylike. About the only resemblance to pea flowers is the fact that one petal, the standard, is larger than the others. In most flowers, the other four petals look very much alike and do not form wings or keel. Each flower has one pistil surrounded by ten or fewer stamens.

The senna subfamily has about 2,200 species of plants, most of them tropical. Most are trees or shrubs, a few are vines or herbs.

The members of this subfamily that have the most butterflylike flowers are the redbuds. Eastern redbud (*Cercis canadensis*) grows in moist woods in eastern United States and in Mexico. The specific epithet *canadensis* is an example of a poorly chosen name. It gives the impression that the eastern redbud is a Canadian plant, but actually the only part of Canada in which it grows wild is a small area along the north shore of Lake Erie.

Redbud is planted as an ornamental tree on lawns because of its beautiful flowers, which clothe the stems in a delicate shade of rosy pink early in spring

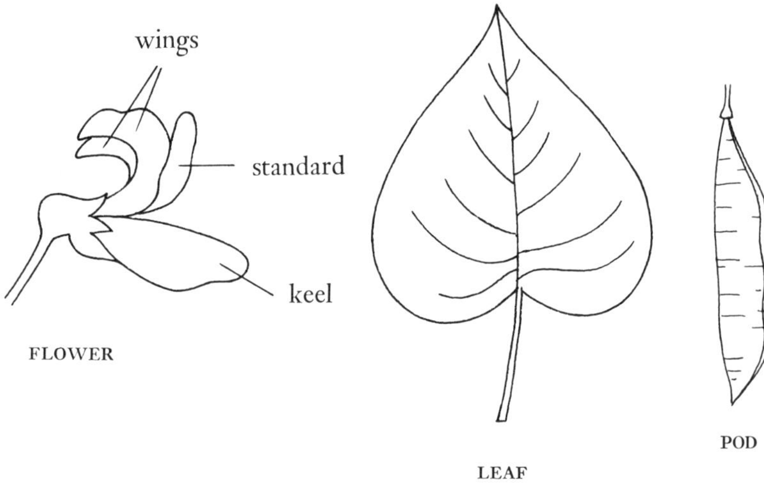

REDBUD

before the leaves appear. If you look closely at a flower you will see that it has two colors; the petals are rose and the sepals purple. Redbud flowers have an acid taste and may be added to salads both for their flavor and their color. Both the flowers and the young pods are good pickled or fried in butter.

A quick glance at a redbud flower could easily give you the impression that redbud is a member of the pea subfamily, but if you examine a flower bud closely you will see one feature that distinguishes the flowers of the pea subfamily from the senna subfamily. While in the bud, the standard of pea subfamily flowers encloses the wings. In the senna subfamily it is the other way around, the wings (or the petals on either side of the standard if there are no clear wings and keel) surround the standard. You usually can see this in mature redbud flowers, too; but in many other members of the senna subfamily you can see it only in the buds; for the petals spread so far apart as the flower opens that no petals enclose any others.

The California redbud (*Cercis occidentalis*) and the Texas redbud (*Cercis texensis*) are similar to the eastern redbud.

Another common name for the American species of redbud is Judas tree, a name originally applied to *Cercis siliquastrum* of the Mediterranean region and Asia. There is a legend that after Judas betrayed Jesus he hanged himself on this tree, and that ever since, the tree has turned red once a year as it blushed with shame because Judas had chosen it. The legend is fictitious, of course. We do not know what species of tree Judas chose (other legends say the fig tree, the poplar, or others); but no matter what species it was, the tree could hardly have a sense of shame.

Another tree associated with Biblical stories is the carob tree (*Ceratonia siliqua*), also called locust, Saint-John's-bread, and algaroba. The very young pods are sour, but as they ripen they develop a sweet pulpy material between the seeds. At this stage they are sometimes eaten by people, but later they become hard and tough and are fed to cattle and swine.

In the New Testament, the food of John the Baptist is said to have been locusts and wild honey; one legend says that the locusts were the pods of the carob tree—hence its other common names of locust and Saint-John's-bread. Another interpretation of this passage from the Bible is that the locusts were actually insects and not plant material at all.

The pods of carob are believed to be the husks mentioned in the parable of the prodigal son, who, after spending all his money, was so hungry that he envied the swine that at least had these to eat.

Whatever the locusts and husks of these stories really were, the brown carob pods are occasionally sold in supermarkets under the name of Saint-John's-bread. Unfortunately, by the time they reach the store they are hard and tough and have little of the sweet pulp. If you open a pod and remove the seeds,

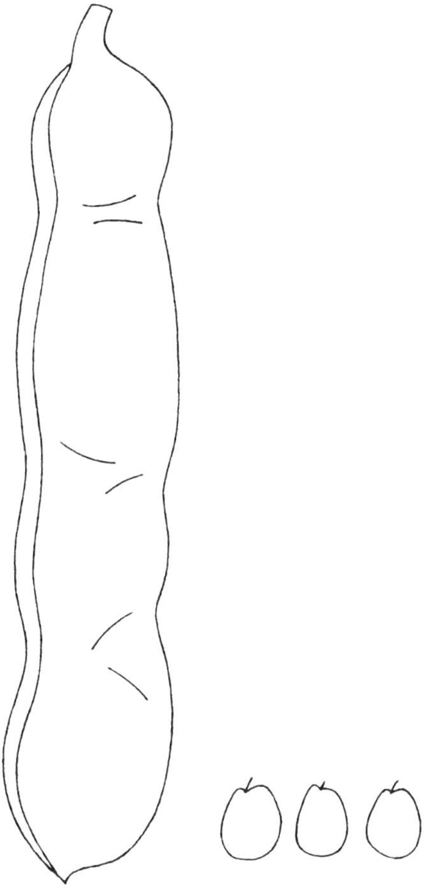

CAROB (*Saint-John's-Bread*)

you will see that they are remarkably alike in size. It is believed that ancient goldsmiths and apothecaries used the seeds as weights and that the seeds were the original "carats."

Health food stores sell carob powder ground from the sweet inner pulp of the pods. Carob powder has a chocolatelike flavor and can be used as a chocolate substitute in some recipes. Carob candy made from carob powder resembles milk chocolate. If you are fond of chocolate, however, you will certainly notice the difference between carob and chocolate.

In the deserts of southwestern United States and Mexico grow two species of tree called paloverde (which means "green stick"), a name that refers to their stems. Both trees drop their leaves almost as soon as they come out, and it is the stems that are

BLUE PALOVERDE

green. One, the blue paloverde (*Cercidium floridum*) has a blue-green bark, and the other, the yellow paloverde (*Cercidium microphyllum*) has a yellow-green bark. In spring, both have yellow flowers somewhat more typical of the senna subfamily than are those of the redbuds; the five petals are nearly alike, and there are no wings or keel. The flowers contain nectar from which bees make a delicious honey.

Paloverde flowers do not last long, but while they do bloom, they brighten the desert with a brilliant blaze of yellow. The Spaniards who explored this part of America in a vain attempt to locate the legendary cities whose streets were said to be paved with gold found only the gold of these flowers, and they called the blue paloverde *lluva de oro,* which means "shower of gold."

The honey locust or three-thorned acacia (*Gleditsia triacanthos*) of eastern United States has small, inconspicuous flowers, and you usually have to look closely to find them. Honey locusts are sometimes planted as ornamental trees along city streets, and if you can find a young tree with branches low enough for you to reach, you might find the greenish flowers about May. Occasionally some flowers have both stamens and a pistil, but usually it is one or the other, although the pistillate flowers may have stamens that produce no pollen. The pistil is bent in a "goose-neck" fashion. The pods are about as conspicuous as the flowers are inconspicuous—espe-

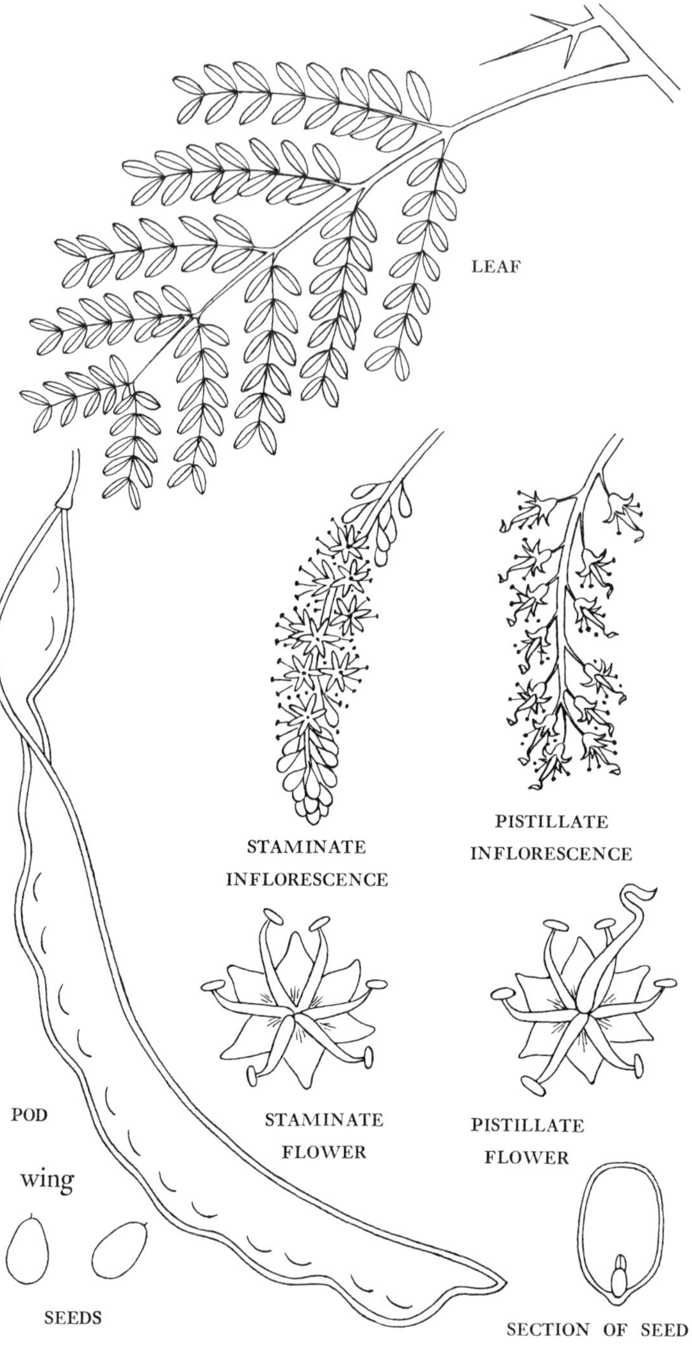

HONEY LOCUST

cially late in the year after the leaves fall. Great clusters of the dark brown pods may hang on the bare trees well into winter, a few falling at a time. The pods do not open. They may be carried about by animals, and they often are blown by wind over the surface of the snow in winter. The pulp between the seeds of the young pods is sweet, and some people enjoy chewing on them for a snack.

A distinctive feature of the honey locust is its hard, woody thorns, which are branched into at least three sharp points and often many more. Some of the thorns are only a fraction of an inch long, others as much as a foot or more. A tree trunk bristling with large, many-pointed thorns is formidable indeed. Before the invention of barbed wire farmers used to make impenetrable fences around their pastures by planting rows of honey locust trees and trimming them to keep them shrubby. The trees planted along city streets usually are of a thornless variety.

The Kentucky coffee tree (*Gymnocladus dioica*), which grows between the Appalachian Mountains and the prairies of the Midwest, has thick, flat, woody pods. Some of the early settlers in Kentucky and Tennessee used the roasted seeds as a coffee substitute. The seeds are poisonous, but the roasting apparently destroyed the poison.

You probably have noticed that some plants are named for the country or the state in which they grow, but there is at least one country that received

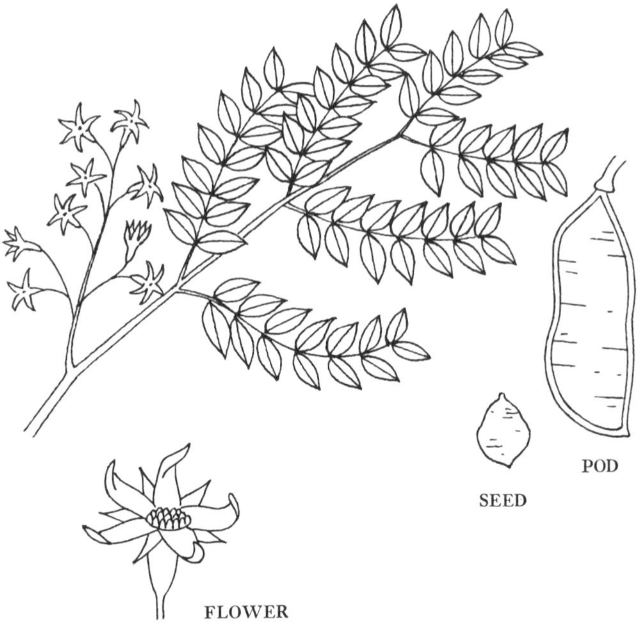

KENTUCKY COFFEE TREE

its name from the product of a plant. During the Middle Ages, Europeans dyed cloth with a fiery red dye called brazil. They obtained it from the wood of a tree called brazilwood or sappanwood (*Caesalpina sappan*). The tree grew in India, which for many years was the only source of the dye. Then, during the early explorations of the Western Hemisphere, a closely related species, *Caesalpina brasiliensis,* was discovered about 1500 in South America. This tree was called brazilwood, too. Its wood yields the same red dye, and soon it was being exported to Europe. So important was this dye that the area in which the trees grew was called Brazil. Today Brazil is the largest country in South America.

Another closely related plant, logwood (*Haematoxylon campechianum*), grows from Mexico to the northern parts of South America. From its wood was obtained a black dye called hematoxylin. This dye is no longer used for dyeing cloth, but biologists still use it for staining microscope slides of plant and animal tissues. It can produce several colors, depending on what other substances are mixed with it. In an acid solution it is red, and in an alkaline (or basic) solution it is blue.

The senna subfamily includes many beautiful ornamental plants, most of them tropical and subtropical. The genus *Bauhinia* has at least 300 species, some of them trees and some giant lianas that grow upward on trees of dark tropical rain forests until they reach sunshine at the treetops sometimes more than 200 feet above the ground.

The orchid tree or mountain ebony (*Bauhinia variegata*) is one of the most popular trees of this genus. Originally from India and China, it is planted extensively in Florida. The purple or lavender flowers usually appear from January to April when the trees are bare of leaves.

Species of *Bauhinia* have simple leaves with two lobes or compound leaves with two leaflets. The twin lobes or leaflets were responsible for the generic name. The plants were named for two sixteenth-century botanists, John and Caspar Bauhin—appar-

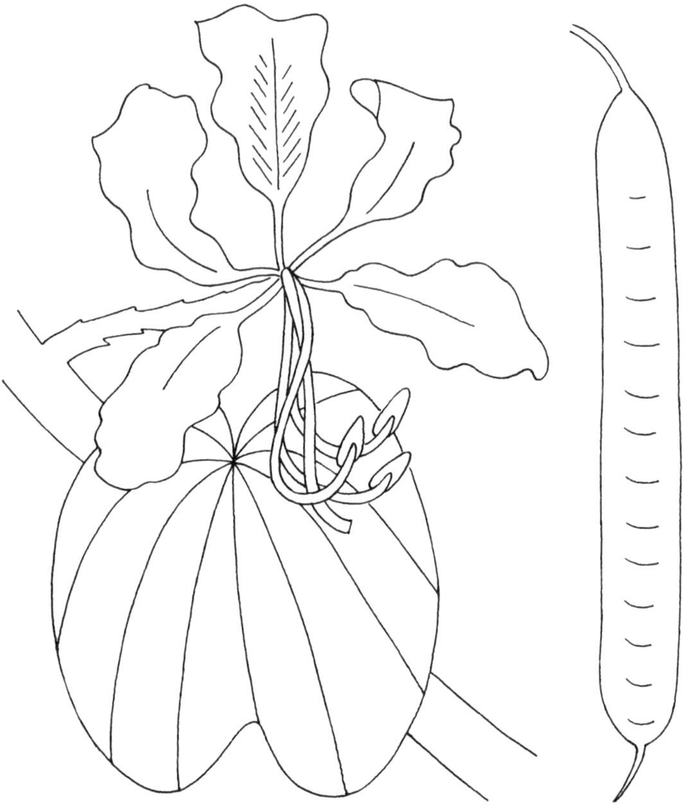

POD

ORCHID TREE

ently in the belief that the two men were twins. Actually, John was nineteen years old when his brother Caspar was born.

Perhaps the most beautiful tree of the senna subfamily is the royal poinciana (*Delonix regia*), also called flamboyant and peacock flower. A native of Madagascar, it is planted in warm countries throughout the world. The spectacular sight of royal poincianas in full bloom with scarlet flowers three or four inches in diameter is said to cause traffic jams as both drivers and passengers pause to enjoy the sight.

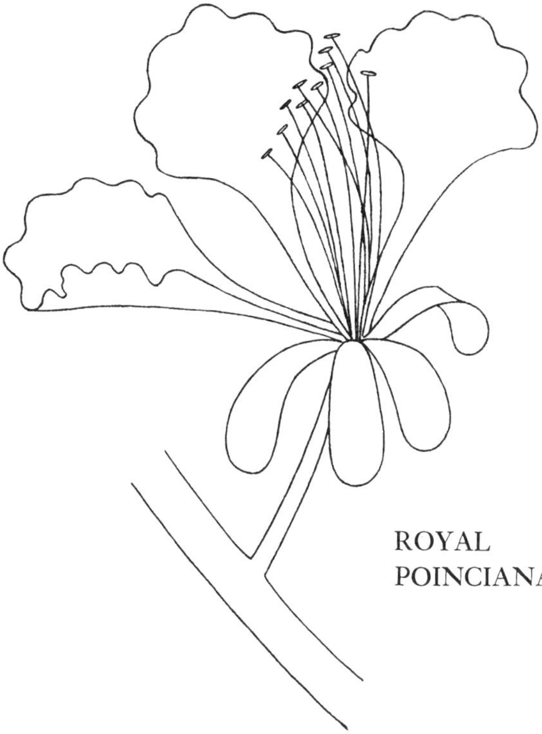

ROYAL
POINCIANA

Chapter 6

The Mimosa Subfamily

(*Subfamily Mimosoideae*)

NEARLY ALL the members of the mimosa subfamily are trees or shrubs, and most of them live in dry tropical and subtropical regions. The plants typically have doubly compound leaves with many small leaflets that give them a ferny appearance. The flowers bear little resemblance to those of the pea subfamily. In the mimosa subfamily the petals of a flower are all alike; there are no standards, no wings, and no keels. The flowers have four or five small sepals and four or five small petals, which you hardly ever notice, partly because the flowers are crowded together in dense inflorescences and partly because each flower has stamens much longer than the petals and sepals. There are at least as many stamens as petals in a flower, and often many more. Frequently the stamens are white or brightly colored, and they

REDHEAD POWDERPUFF

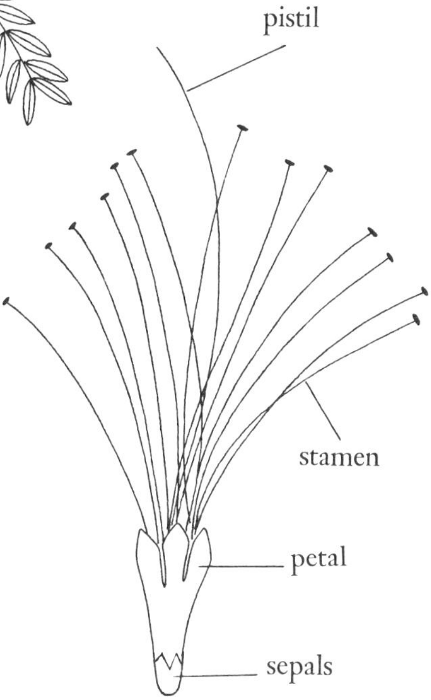

FLOWER

give a head the appearance of a powderpuff or pompom. In fact, one shrub from Bolivia, *Calliandra hematocephala,* has crimson stamens and is called redhead powderpuff. The flowers have one pistil, which ripens into a pod similar to those of the pea and senna subfamilies.

Mesquite (*Prosopis juliflora*) has been called the western honey locust, Texas ironwood, and honeypod. It is a small tree with a long taproot that may grow downward in the soil to depths of 60 feet or more. This enables it to grow in the deserts of southwestern United States and Mexico, where, for most

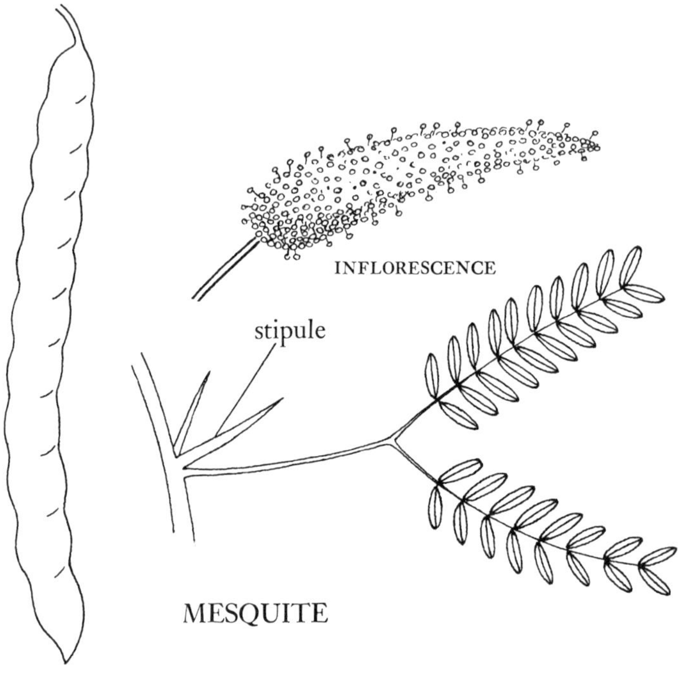

MESQUITE

of the year, the only water may be deep underground. Where roots must grow to great depths before they reach water, mesquite does not grow as tall as it does where the water supply is nearer the surface. The local people claim that they can estimate the distance down to the level of ground water by the heights of mesquite trees—the taller the trees, the nearer the water is to the surface.

Almost every part of the mesquite has some use. The fragrant flowers, which are densely clustered in racemes, secrete nectar. Bees visit them to gather the nectar from which they make a delicious honey, and lizards and squirrels come to eat the sweet flowers. The first flowers bloom in April, and new flowers continue to appear well into summer.

In July the pods are ripe. They have a high sugar content, and the Indians used them for food and fermented them to produce a mild beer. Livestock enjoy the pods, too, and goats will even climb into the trees to reach them. Unfortunately, the goats also eat the leaves and twigs, and if they eat enough of them, the trees are killed.

Mesquite wood is not very strong, but like black locust wood, it does not rot when in contact with the soil and so has been used for fence posts and railroad ties. In pioneer days the hubs and spokes of wagon wheels were made from mesquite wood, and the Navaho Indians used it for their bows. The wood burns slowly and gives off a great deal of heat, and so it makes a good campfire.

The Indians collected the gum that exudes from wounded mesquite trunks and used it for mending pottery. They wove baskets from the twigs and obtained a dye from the sap. Even the babies had diapers made from mesquite bark, beaten and rubbed to make it soft.

Small animals often find refuge in the thorny branches of mesquite, for they can run safely among the thorns, while a large pursuer is more likely to become scratched or cut by them. The spreading branches cast a dense shade most welcome to travelers on a hot summer day.

About the only unfavorable thing said about the mesquite is that it has spread from its native desert areas of the Southwest eastward into the grasslands of Texas, Oklahoma, Missouri, and Louisiana, thus ruining grazing land on which cattlemen depend. The seeds, however, are spread by the cattle themselves. When the animals eat the pods, the seeds remain undamaged in their digestive tracts and are passed in their excrement. If the cattle should be moved from a desert area where mesquite is in fruit to a grassland during the time the seeds are inside them, the seeds, of course, will be deposited on the grassland. There is some evidence that mesquite seedlings flourish only in overgrazed grassland and not in grassland that is well managed. So perhaps the invasion of grasslands by mesquite is not entirely the fault of the mesquite.

The screwbean mesquite (*Prosopis pubescens*),

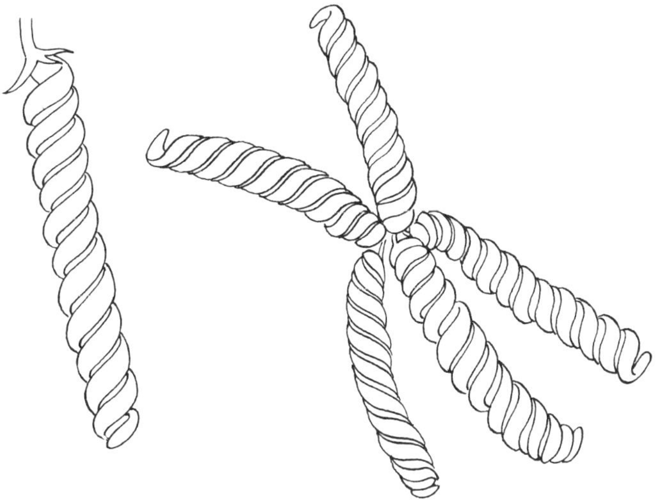

SCREWBEAN MESQUITE PODS

also called screwpod mesquite and toronillo, is similar to mesquite, but the pods are twisted into tight spirals, some having as many as twenty turns.

The mimosa subfamily can be something of a headache when it comes to common names. Three of its genera are *Acacia*, *Albizzia*, and *Mimosa*. Species of *Acacia* that grow in Australia are called wattle, those that grow in central Africa are called thorn trees, those that grow in southwestern United States are called acacia, and the species sold by American florists are called mimosas. *Albizzia julibrissin* has several common names including both acacia and mimosa, though many persons call it silk tree. Recall that species of *Robinia* (subfamily Papilionoideae) have been called acacias, too, and so has *Gleditsia tricanthos* (subfamily Caesalpinioideae). Just to prevent confusion, for the rest of this chapter, whenever the common name acacia is used, it will refer to a species of *Acacia,* and mimosa will refer to a species of *Mimosa.* Some other common names will be used where they are appropriate, but at least we will not call a species of *Acacia* mimosa.

The acacias are mostly trees and shrubs, and most of them are native in Australia. There they are called wattles because settlers used them to build their houses by the "wattle and daub" method. A wattle is a structure formed of interwoven branches, and the settlers used the stems of these plants and then daubed mud over them to complete the walls.

Acacia stems usually have thorns—hence their common name of thorn tree in Africa. You probably have seen these thorn trees in pictures of the wildlife of the African savannahs. A savannah is a grassland with widely-spaced trees; most of the trees in these savannahs are thorn trees. In several species swellings called galls form in the thorns. Ants sometimes hollow out holes in the galls and use them as homes. Baboons bite open the galls of sickle-lobe acacia (*Acacia drepanolobium*) and eat the ants in them.

The adhesives on the backs of some stamps is made from gum arabic (or gum acacia), a gum that comes from the wounded trunks of several species of *Acacia* from Africa and India.

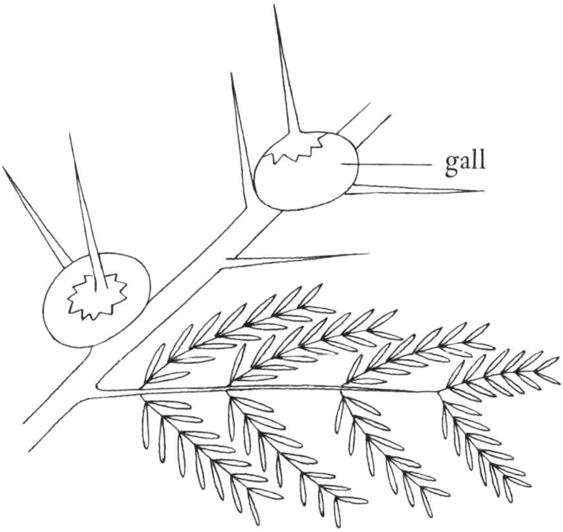

SICKLE-LOBE ACACIA

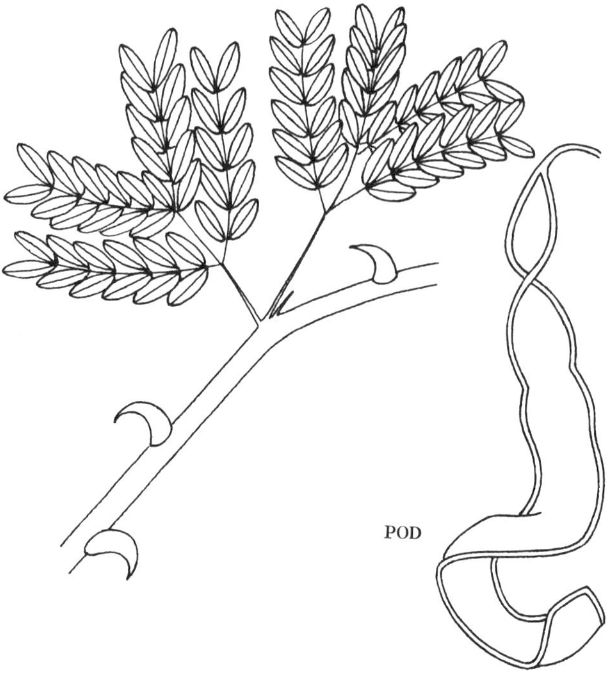

CATCLAW ACACIA

In the southwestern deserts of the United States grow several acacias, among them the catclaw acacia (*Acacia greggii*). Its thorns are about the size of a cat's claws and are curved like them, too. They inflict serious cuts on anyone who becomes entangled in the branches. If we are to judge from some additional names of this tree, people have had unpleasant experiences with it: devil's claws, tear-blanket, and wait-a-minute.

One of the most interesting members of the mimosa subfamily is the sensitive plant (*Mimosa pudica*) of the tropical parts of South America. This small, shrubby plant is frequently grown as a curiosity in greenhouses in colder climates. It has also been a subject of interest to botanists, for the leaves are sensitive to touch.

Each leaf is doubly compound. It has four large leaflets palmately arranged, and each of these is divided into many smaller leaflets that are pinnately arranged. The leaves, like those of some other leguminous plants, show sleep movements, but they also respond similarly to touch. During daylight hours, if a plant has been undisturbed for some time, the leaflets all lie in a horizontal or nearly horizontal position. At night, the small leaflets bend upward until the upper surfaces of opposite leaflets meet, and the entire leaf droops. The next morning the leaf and its leaflets return to their original position.

A very light touch on one of the small leaflets of an undisturbed plant may cause only it and the leaflet opposite it to bend upwards. A firmer touch (or pinching a small leaflet) usually causes several leaflets to respond, first the pair of leaflets which was touched, then the adjacent pair, then the next pair, and so on until perhaps all the small leaflets on one of the larger leaflets have folded up. A firm slap usually causes the entire leaf to droop and all its leaflets to fold up. If the slap is severe enough, other leaves will respond in the same way. The first leaf

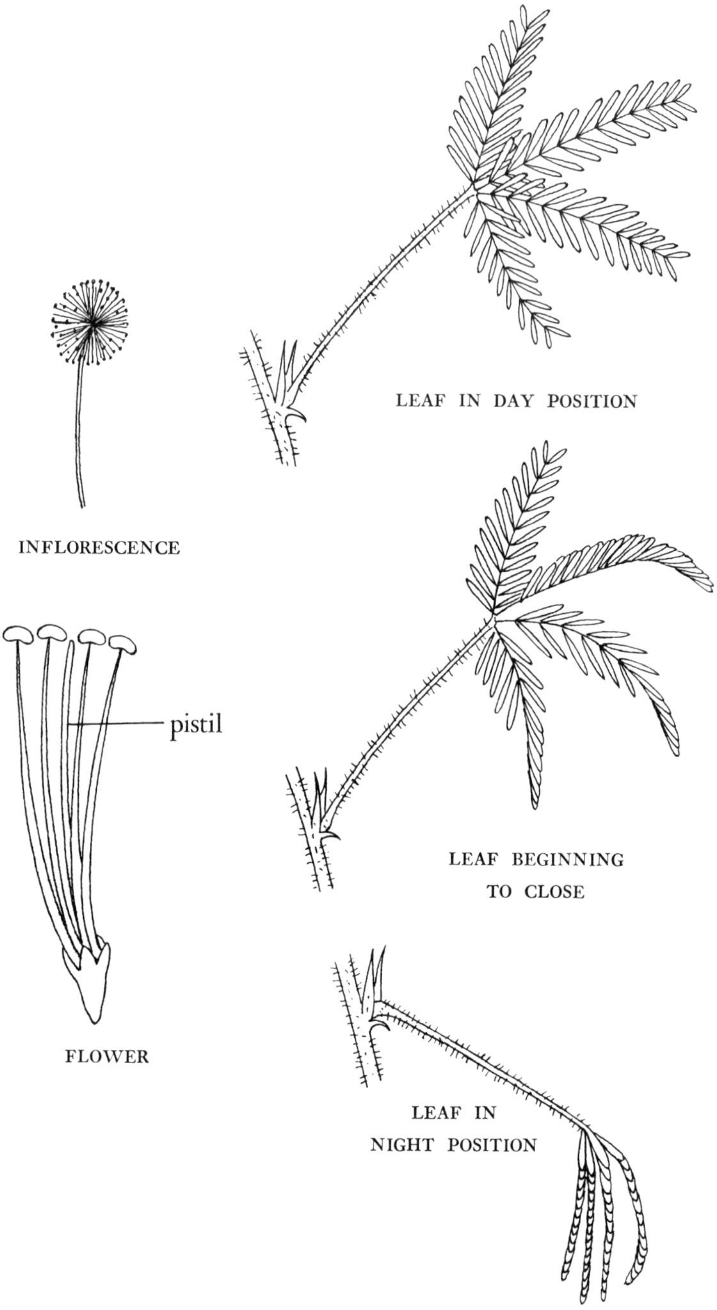

SENSITIVE PLANT

to respond is the slapped one, then those nearest to it, then those farther away. Usually it takes less than a minute for the entire plant to respond.

A violent shock can cause all the leaves to respond immediately. Shaking a sensitive plant or moving it from one place to another in any but the gentlest manner can cause the entire plant to respond. If left undisturbed, a "shocked" plant recovers in a few minutes or an hour or so. The more severe the shock the longer it takes the plant to recover.

Conclusion

BECAUSE of their high protein content, beans and peas have been some of the most important foods of mankind. The only plants of equal or greater importance are the cereal grains such as wheat, rice, and corn. Both types of food can be harvested when they ripen and then can be stored dry for long periods of time without spoiling.

This was very important in the beginnings of civilization when there were no conveniences like canning and freezing. It meant that people who raised these foods could keep a supply of food over winter or other unfavorable periods, and so they could stay in one place and not have to travel to areas with more favorable weather to look for food. Remaining permanently in one place, they formed villages and later cities.

Proteins are composed of amino acids, some of which are essential in our diets and some of which are not. The proteins in most meats and other animal products contain all the essential amino acids in just about the proportions we need them. Plants, on the other hand, lack several of the essential amino acids or have them in such low concentrations that they cannot supply our needs. Even the protein-rich leguminous plants do not provide all the amino acids we need. Fortunately, the amino acid content of beans and peas and the amino acid content of whole-grain cereals (which are fairly rich in protein, but not so much as leguminous plants) nearly balance each other. If our meals contain both whole-grain cereals and leguminous plants, we can obtain a nearly balanced diet with respect to amino acids. It is possible, then, to rely on leguminous plants for most of our protein if we supplement it with protein from whole-grain cereals—and occasionally some meat, fish, milk, eggs, or cheese.

Each of the major ancient civilizations raised at least one cereal grain and one leguminous plant that were important items of their diets. In much of the Far East, for instance, people ate rice and soybeans, and in India they ate rice and garbanzos or chick peas (*Cicer arientum*). Wheat was the most important cereal grain in the Middle East and the Mediterranean area, but barley, oats, and rye were eaten there as well; lentils, broad beans, and peas were the important leguminous plants. Corn was the cereal grain of the American Indians, who also raised kidney beans and lima beans.

Today, as the world is becoming more and more overpopulated, it is no longer possible to raise enough meat or to catch enough fish to provide everyone with protein for a good diet. Already, poor people all over the world use leguminous plants as

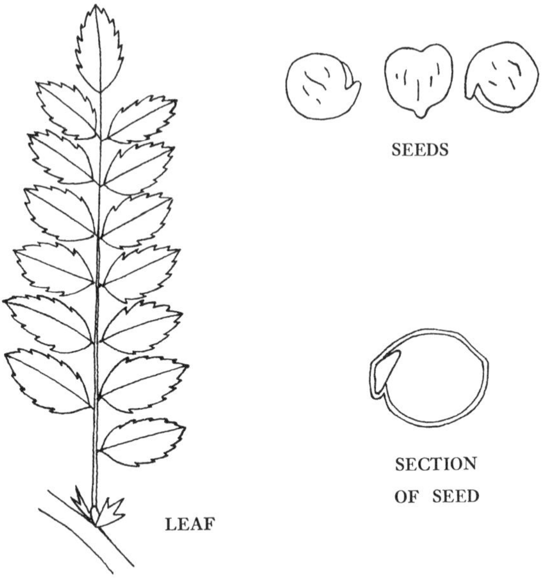

GARBANZO
(Chick Pea)

their major source of protein. It seems likely that as the human population grows larger, more people will find it too expensive to eat meat regularly and will have to obtain most of their protein from beans and peas.

Although people have used many kinds of plants for many different purposes throughout history, the leguminous plants together with the cereal grains, more than any other plants, made the origin of civilization possible. Civilization still depends on them today, and probably it will continue to do so for a long time into the future.

Appendix

Meanings of specific epithets used in this book:

alba: white
arientum: resembling the head of a ram
aureus: golden
brasiliensis: from Brazil
campechianum: from the Bay of Campeche, in southeastern Mexico, where *Haematoxylon campechianum* was first discovered
canadensis: Canadian
canescens: grayish and covered by soft hairs
castorea: of beavers
coccineus: scarlet
corniculatus: horned
crista-galli: cockscomb
culinaris: pertaining to food or the kitchen
dioica: having staminate and pistillate flowers on separate plants
drepanolobium: sickle-lobed
esculenta: edible
faba: bean
floribunda: flowering abundantly

floridum: flowering

foenum-graecum: Greek hay

glabra: smooth

grandiflora: having large flowers

greggii: named for Josiah Gregg, a frontier trader who traveled in the American southwest

hematocephala: red head

herbacea: herbaceous (not woody)

hispida: covered with stiff hairs or bristles

hybridum: hybrid

hypogaea: underground

incarnatum: flesh-colored

julibrissin: ancient Persian name of *Albizzia julibrissin*

juliflora: having flowers in dense, cylindrical inflorescences (racemes)

latifolius: broad-leaved

lepidota: scaly

lobata: lobed

lunatus: moonlike

macrobotrys: a large cluster of grapes (the large, spherical pods of *Strongylodon macrobotrys* hang in groups resembling gigantic clusters of grapes)

max: the Spanish name for the mung bean (the soybean was named *Glycine max* in the mistaken belief that it was the same as the mung bean)

metcalfei: named for J. K. Metcalfe, who introduced *Phaseolus metcalfei* into the American Southwest

microphylum: having small leaves

neo-mexicana: New Mexican
occidentalis: Western, occidental
odoratus: fragrant
officinalis: medicinal
phaseoloides: like *Phaseolus* (the genus that includes kidney beans, lima beans, and some other beans)
pratense: growing in meadows
precatorius: prayerful, praying
pseudoacacia: false acacia
pubescens: downy, hairy
pudica: modest, bashful
purpureum: purple
regia: royal
repens: creeping
sappan: from *sapang*, the Malay word for sappanwood
sativa, sativum: cultivated
scoparius: broom
sepium: growing on hedges or used for hedges
sesquipedalis: a foot and a half long
siliqua, siliquastrum: resembling a silique (a fruit typical of the mustard family)
sinensis: Chinese
spinosa: spiny
subcarnosus: rather fleshy
subvillosus: slightly hairy
tesota: from the Spanish word *tieso*, which means "stiff"
texensis: Texan

tinctoria: used for dyeing
triacanthos: three-spined
tuberosa: having tubers
varia: various, changeable
variegata: variegated
viscosa: viscid, sticky
vulgaris: common

Index